T0123260

How to Select the Right Centrifugal Pump

A Brief Survey of Centrifugal Pump Selection Best Practices

Robert X. Perez

authorHOUSE®

AuthorHouse™
1663 Liberty Drive
Bloomington, IN 47403
www.authorhouse.com
Phone: 1 (800) 839-8640

Published by AuthorHouse 09/23/2015

ISBN: 978-1-5049-2267-8 (sc)
ISBN: 978-1-5049-2268-5 (hc)
ISBN: 978-1-5049-2266-1 (e)

Library of Congress Control Number: 2015911270

Print information available on the last page.

Table of Contents

Dedication

To my mother, Dolores Perez, for teaching me the importance of education and hard work.

Acknowledgement

The author would like to thank:

1) His wife, Elaine, for providing valuable feedback on the layout and format of the book and helping with editing.
2) Jacques Chaurette, Mechanical Engineer, for providing the material found iin Appendix A: "How to Calculate Pipe Friction Head Losses for Newtonian Fluids"
3) Julien LeBleu of Sage Technologies LLC for reviewing the manuscript and pointing out areas for improvement.

Preface

The goal of this short book is to provide those who need to select reliable and efficient centrifugal pumps for industrial applications with some practical tips and advice on how to evaluate the many pump design possibilities available on the market today.

The technical discussion presented here assumes that the reader has a basic understanding of centrifugal pump construction and terminology. My book, "Operators Guide to Centrifugal Pumps" is an excellent primer for those who feel the need for a basic review of centrifugal pump technology before tackling the topics in this book.

Choosing a centrifugal pump from the countless options available can be daunting, but someone has to make the decision. Many factors, such as the required flow, differential pressure, suction conditions, etc., must be weighed against the capital costs and cost of energy for the pumps considered. To determine the right pump, you must consider the overall cost of ownership, which includes capital cost, operating costs, and maintenance cost. What good is a low cost pump if it is inefficient or if is costly to maintain?

The selection methodology offered in this book focuses mainly on hydraulic design considerations, but also touches on mechanical design details. A large part of the book is devoted to key hydraulic deign parameters, such as specific speed (Ns) and suction specific speed (Nss), NSPHa, NPSHr, and NPSH margin ratio. Analyzing these basic hydraulic parameters allow you to quickly determine if a centrifugal pump makes sense for your particular application.

If you do decide a centrifugal pump will work for your application, then you need to be able to evaluate the various bids returned by pump manufacturers. Chapter 5 covers how to tabulate quotes from pump manufacturers in order to properly evaluate their bids and select the best overall option.

Using the advice in this book along with your company's centrifugal pump guidelines, should ensure the selection of the optimum pump for your given application. Remember when in doubt, your company's specifications along with current governing pertinent standards, such as API, PIP, ANSI, etc., should be considered the final word in the selection process.

I hope readers will find this book to be a useful addition to their technical libraries. My ultimate goal was to provide a valuable reference book that will be relevant for years to come.

Robert X. Perez

Chapter 1
Introduction to Centrifugal Pumps

Pumps are required throughout the process industry to move liquids from point A to point B, as shown in Figure 1.1 below where a pump takes suction flow from an overhead process vessel and pumps it into the process. Pump functions may involve moving liquid from a lower pressure to a higher pressure, lifting the liquid from a lower to a higher elevation or a combination of both. Liquid normally wants to move from a higher pressure to a lower pressure and from a higher level to a lower level. Getting liquid to move against its normal tendencies requires the addition of energy. That's where pumps come in. Pumps are devices that transform mechanical energy into useful liquid energy.

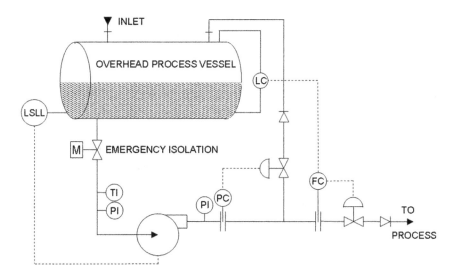

Fig. 1.1 - A centrifugal pump takes suction from an overhead process vessel and pumps product into a process.

Centrifugal pumps are one of the simplest of all the pump designs. They have one moving part. The rotor has an impeller that accelerates liquid from its suction eye or inlet (see Figure 1.2) to a maximum speed at its outer diameter.

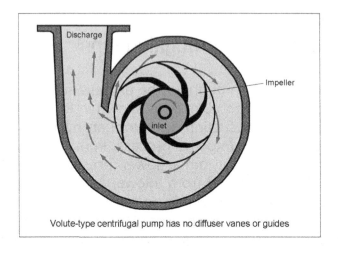

Volute-type centrifugal pump has no diffuser vanes or guides

Figure 1.2- Centrifugal Pump Impeller and Volute

The liquid is then gradually decelerated to a much lower velocity in the stationary casing, called a volute casing. As it slows down, due to the increasing cross sectional area of the casing, pressure is developed, until full pressure is developed at the pump's discharge. This simplicity of design and operation is what makes centrifugal pumps one of the most reliable of pump designs, assuming they are applied properly.

This process of converting velocity to pressure is similar to holding your hand outside of a moving automobile. As the high velocity air hits you, it slows down and pushes your hand back due to the pressure developed. Similarly, if you could insert your hand into the pump casing at the impeller exit and "catch" the liquid, you would feel the pressure produced by dynamic action of the impeller. When any high velocity stream slows down, pressure is created. (This effect is called Bernoulli's Principle, which simply states that energy is always conserved in a fluid stream.) The greater that the impeller diameter or rpm is, the greater the exit velocity is and, therefore, the higher pressure developed at the pump's discharge.

Another benefit of centrifugal pumps is that they can cover a wide range of hydraulic requirements. Thus, they can be used in a wide range of flow and pressure applications. They can easily provide flows from less than 10 gallons per minute (37.9 liters per minute) to well over 10,000 gallons per minute (gpm) (379 000 liters/minute). Centrifugal pump impellers can easily be staged, that is, arranged so that one impeller's output is directed to a subsequent impeller; over 4500 pounds per square inch (psi) (31026 kPa) of pressure can be generated.

One key disadvantage of centrifugal pumps is that their efficiencies are usually less than that of positive displacement pumps. Whereas positive displacement pumps can deliver efficiencies greater than 90%, centrifugal pump efficiencies can range from less than 30% to over 80%, depending on the type and size. Here are a few samples of centrifugal pump efficiencies:

1. 10,000 gpm centrifugal pump—75 to 89%
2. 500 gpm centrifugal pump—60 to 75%
3. 100 gpm magnetic drive pump—40%
4. 100 gpm canned motor pump—35%

This is the price you pay for simplicity.

Head versus Pressure

Before we can talk about centrifugal pump performance, we must understand the difference between pressure, which is usually given in pounds per square inch (psi) and liquid head, which is usually given in feet, inches, or mm of liquid. In Figure 1.3, we have three pressures gauges connected to the bottom of three different 115.5-feet (35.2 m) liquid columns. Notice that pressure gauges read, from left to right, 60 psi (414 kPa) on the liquid with a specific gravity (S.G.) of 1.2, 50psi (345 kPa) on the liquid with a specific gravity of 1.0, and 35 psi (241 kPa) on the liquid with a specific gravity of 0.70. From these observations, we can infer that, given a fixed column height, the denser or heavier the liquid in a column, the greater the expected reading at the pressure gauge. Conversely, we can state with confidence that a lighter liquid will result in a proportionally lower reading

at the pressure gauge. The formulas used to determine exactly
how pressure relates to head and vice-versa are given below.

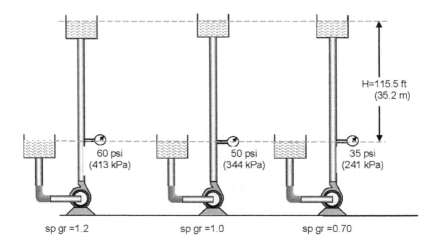

Figure 1.3 - Pressure gauges at the bottom
of three different liquid columns

$$P(psi) = \frac{(H\,in\,ft)(S.G.)(\gamma_{water}\,in\,\frac{lb}{ft^3})}{(12\frac{in}{ft})^2} = \frac{(H\,in\,ft)(S.G.)(62.4\frac{lb}{ft^3})}{144\frac{in^2}{ft^2}} = \frac{(H\,in\,ft)(S.G.)}{2.31}$$

$$P(kPa) = \frac{(H\,in\,m)(S.G.)(\rho_{water}\,in\,\frac{kg}{m^3})(g\,in\,\frac{m}{s^2})}{(1000\frac{Pa}{kPa})} = \frac{(H\,in\,m)(S.G.)(1000\frac{kg}{m^3})(9.81\frac{m}{s^2})}{1000\frac{Pa}{kPa}} = 9.81(H\,in\,m)(S.G.)$$

Where *SG* is the specific gravity of the fluid.

To use these formulas, you need to understand the liquid
property specific gravity, which is simply the ratio of the liquid's
density to the density of water (62.4 lbs/ft³ or 1000 kg/m³) at
a given temperature. The denser the liquid is, the higher its
specific gravity will be.

Let's go through a few simple examples.

1. First, we will assume we have a 100-foot column of water.
    Water has a specific gravity of 1.00, so the pressure you
    would expect at the bottom of the column to be (100 x
    1)/2.31 or about 43.3 psi.

2. Next, we will assume we have a 100-foot column of gasoline, which has a specific gravity of 0.75. (This means one cubic foot of gasoline weighs 75 % the weight of one cubic foot of water.) We can expect the pressure gauge to read (100 x 0.75)/2.31 or 32.5 psi.
3. Finally, we will assume we have a 100-foot column of brine, which has a specific gravity of 1.2. (This means one cubic foot of brine weighs 120 % the weight of one cubic foot of water.) We can expect the pressure gauge to read (100 x 1.2)/2.31 or 51.94 psi.

If the specific gravity, or density, of the liquid is known, we can readily convert pressure head into pressure. Although pressure and head are not identical quantities, they are closely related and are often used interchangeably.

Head is a convenient form of stating pressure in centrifugal pump applications. Most manufacturers test their pumps with water, so you need to be able to determine how your pump will perform on the liquid you are pumping. If a pump is capable of delivering 200 feet (61 m) of head, you can conclude your pump will add 200 feet (61 m) of liquid head to whatever pressure you have on the suction of the pump, regardless of the specific gravity of the liquid being pumped. Let's also say you are pumping gasoline and that you have 50 feet of suction pressure. You can now calculate the discharge pressure of the pump. (Here we will assume there are no pressure losses in the pump's piping.)

$$P_d = \frac{\left(H_s + H_{Pump}\right) \times S.G.}{2.31}$$

In the equation above, P_d is the Discharge Pressure, H_s is the suction head in feet, H_{pump} is the pump's head rise in feet, and S.G. is the specific gravity. Plugging in the values provided we get:

$$P_d = \frac{(50+200)\times 0.75}{2.31} = 81.2\,psig$$

It is important to note that, in our example, the pump's discharge pressure (P_d) is in gauge pressures (psig) as opposed to an absolute because we chose to state SH in feet of liquid above atmospheric pressure.

Another way of looking at pump head is to imagine what would happen if you opened a bleeder near the pump's discharge nozzle to see how high the pumping liquid would shoot up due to the generated head. Using the example above, where the pump is adding 200 feet of head to 50 feet of suction pressure above atmospheric pressure, you should expect to see a liquid stream soar 250 feet into air, excluding valve losses and air resistance.

Centrifugal Pump Performance

Table 1.1 shows a listing of typical performance data for a centrifugal pump. You will notice the table lists "head" (a form of pressure that the pump produces), "NPSHr" (the amount of suction head over the vapor pressure required for normal operation), the required horsepower, and the pump efficiency for five different flow conditions. This is typical of pump test data. The pump to be tested is placed on a well-instrumented test stand so that these performance values can be determined. From the table below, you can readily see that at 400 gpm you can expect 195 feet of pressure head. In addition, 32.8 horsepower will be required to operate the pump at this flow condition. Similarly, at 800 gpm, you can expect only 120 feet of pressure, and 48.5 horsepower will be required to operate the pump at this flow. Similar data is obtained and is used in the metric system but flow is in liters per minute, head is in meters, NPSH is in m, and power is in kilowatts.

Table 1.1 Typical Centrifugal Pump Performance Data

Flow (gpm)	Head (ft)	NPSH (ft)	HP
0	205	4	15.0
200	205	5	25.9
400	195	6	32.8
500	185	7	37.7
600	170	8	42.9
800	120	12	48.5

Pump manufacturers usually convert this tabular performance data into graphical formats called pump performance curves. These curves provide pump users a means of visualizing the performance data and allowing them to quickly determine how changes in flow can affect pressure and horsepower. If you look at Figure 1.4, you will see pump performance curves generated from our pump performance data in Table 1.1.

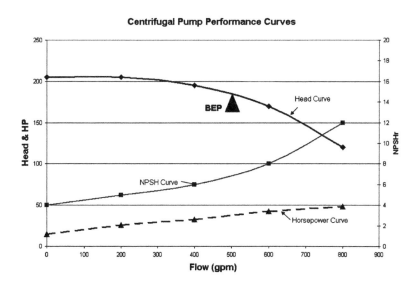

Figure 1.4- Typical Pump Curve

You can see that:

- Pump head is fairly constant until the flow increases above 400 gpm (approximately) and then drops rapidly. The flatness of the head curve at lower flows is why we call centrifugal pumps constant head devices.
- NPSHr, or suction requirement, increases steadily as flow increases.
- Horsepower also increases as flow increases. (It is important to note that not all pump curves have a continuously rising horsepower curve like this one. In general we can say that radial flow pumps have continuously rising horsepower versus flow curves, mixed flow pumps have relatively flat horsepower versus flow curves, and axial pumps have continuously falling horsepower versus flow curve.)

This visual display of pump performance data is a useful trouble shooting and analysis tool.

Basic Centrifugal Pump Construction

Most centrifugal pumps have the following common elements (see Figure 1.5):

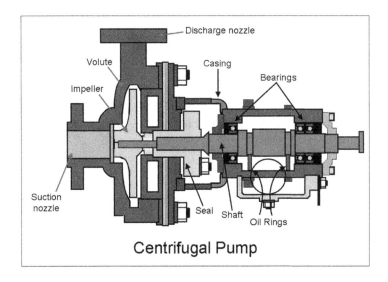

Figure 1.5- Centrifugal Pump Cross Section

1. Impeller, which adds energy to the liquid by accelerating it
2. Pump casing and volute, which contains the liquid being pumped and decelerates the liquid expelled by the impeller
3. Shaft seal, which allows rotation of the rotor while preventing product leakage around the shaft. Mechanical seals are commonly used in process applications
4. Shaft and bearings, which maintain the position of the rotor with respect to the pump casing

Mechanical seal and bearing failures tend to be the most common component failures within centrifugal pumps. High vibration, shaft deflection, high operating temperatures, chemical attack, and the build-up of solids can cause mechanical seals to fail prematurely, while bearing life can be curtailed by excessive shaft vibration and lubricant contamination. Extra attention should always be given to mechanical seals, bearings, and their corresponding support systems to ensure they are properly designed and applied. Once the designs of these key elements have been optimized for the given application, detailed operating and maintenance procedures should be developed and then faithfully followed to ensure reliable and economical pump operation.

Types of Centrifugal Pumps

Pumps come in all shapes and sizes to fit the hydraulic requirement of the industry. Here are just a few common designs:

Fig. 1.6 - Close-coupled centrifugal pump: A close coupled centrifugal pump is directly coupled to an electronic motor, thereby reducing the space requirement. However, this design requires that the electric motor be removed to replace the mechanical seal.

Fig. 1.7 – Back pull-out, overhung centrifugal pump: This is a centrifugal pump design with a flexible coupling between the electric motor and the pump. This design tends to take up more space but allows the removal of the impeller, bearing housing and seal for maintenance by removing the coupling spacer and unbolting the rear of the pump.

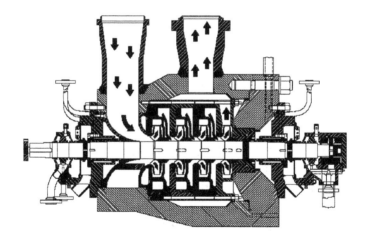

Fig. 1.8- Multistage centrifugal pump: By installing multiple impeller stages in series, a centrifugal pump can generate higher differential pressures. This pump is using four impellers in series. The impeller on the far left is the first stage and the impeller on the far right is the final impeller stage.

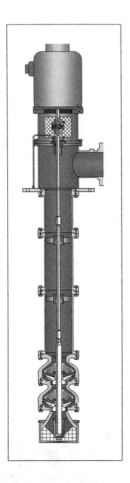

Fig. 1.9 - Vertical, multistage centrifugal pump: These pumps are typically used in sumps to ensure there is adequate submersion depth below the liquid level. Vertical multistage pumps can also be used inside a pressure containing can to increase the suction head available to the first stage impeller in order to avoid cavitation.

Fixed Versus Variable Speed Operation

There are two design choices when specifying centrifugal pumps. You can either opt for a fixed or constant speed application or a variable speed application. Variable speed operation allows the user to vary the developed pump head by varying the pump speed to meet the process requirements. As the pump speed increases its developed head increases, therefore increasing the flow and the horsepower demand (see Fig. 1.10).

The simplest installation and therefore the lowest cost design option tends to be a fixed speed pump driven by a constant speed electric motor. In fixed speed applications, control valves are often used to modulate the pump flow. By opening or closing a control valve, the pump flow is affected by increasing or decreasing the back-pressure on the pump. Make sure to size the control so that it can handle 110% to 125% of the rated pump flow, for upset conditions.

In contrast, variable speed applications are more expensive options due to their inherent complexity. Variable speed pumps can be driven by electric motors with variable frequency drives, steam turbines, or gas turbines. Typically the potential cost savings realized by variable speed operation justifies the additional cost of variable frequency drives. Steam turbine drives are usually selected whenever there is an excess supply of steam and gas turbines tend to be used if readily available fuel sources are more convenient than electricity.

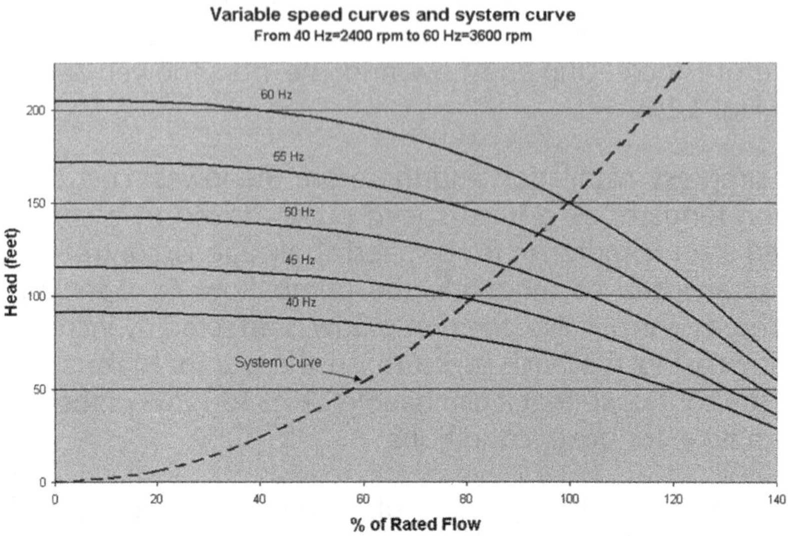

Variable speed curves and system curve
From 40 Hz=2400 rpm to 60 Hz=3600 rpm

Fig. 1.10 - As a pump speed increases its developed
head increases, therefore increasing the flow.

In this book, we will concentrate on fixed speed application; however all the concepts presented here relate to variable speed applications. A good practice is to select your pump based on maximum or rated speed condition. Since the highest energy conditions are seen at the maximum speed, it makes sense to ensure the pump you select fits best at your highest process demands.

Pumping Systems

When multiple pumps are piped together, as shown in figure 1.11, we create pumping systems that offer process facilities and pipelines multiple capabilities. Pumps can either be piped in series (in a line) or in parallel (side by side). When pumps are piped in series their pressures add together. When pumps piped in a parallel arrangement, their flows add together. In parallel, as each additional pump is turned on, more and more flow can be realized by the system. Keep in mind that pump and piping systems must be carefully designed to adequately handle the overall system requirements. Every aspect of a piping system

14

should be considered to ensure that the production goals can be realized.

Fig.1.11 - Pumping system

The Importance of System Head Curve

Every pump manufacturer would like to supply the perfect pump for every application they quote. However to do this, the manufacturer requires the future pump user to provide an accurate system head curve that describes the capacity and head needed for your operating conditions. In this section, we will define what a system head curve is, why it is important, and how it is generated.

A system-head curve is an analytical or graphical representation of the relationship between the flow and the hydraulic head requirements of a given piping system. Hydraulic head requirements are related to the suction head, the discharge head, and the hydraulic head losses of the piping system. Since hydraulic head requirements and losses are functions of the flowrate, size and length of pipe, and size, number and type of fittings, each system head curve has a unique shape.

Normally, at least one point on the system curve is given to the pump manufacturer in order to help him select the pump properly. However, it is highly desirable to graphically superimpose the entire system curve over the head-capacity curve of the pump in order understand the interactions between the pump performance curve and the system head curve. By definition, the intersection of the pump performance curve with the system-head curve defines the operating point of the pump and piping system. Once the system curve is defined, we can plot various pump curves on top of the system curve and hopefully select one that matches the process needs. Without this system curve, there is not much of a chance of coming up with the right pump.

To create a system curve we must first plot the desired capacities against the required head over the total anticipated operating range of the pump. The head will be measured in feet or meters and the capacity will be measured in gallons per minute or cubic meters per hour. The general equation for a system curve is given by the following equation:

$$H_{system} = (P_{downstream} - P_{upstream}) + \text{Elevation} + \text{All line losses}$$

Putting this equation in words, we would say that the system head is the sum of 1) the difference between the downstream pressure and the upstream pressure, 2) the elevation difference between the downstream liquid level and the upstream liquid level, and 3) all the line losses. For the remainder of the chapter we will use the following shorthand version of the equation:

$$H_{system} = (P_{ds} - P_{us}) + H_{static} + L$$

H_{static} is defined as the vertical distance from the surface of the liquid in the suction tank or pit to the level of the liquid level in the discharge tank. H_{static} can be positive or negative.

As the liquid flows through the piping and fittings, it is subject to the friction caused by the internal finish of the piping, restrictive passages in the piping fittings and any additional hardware that has been installed in the system. The resulting "pressure drop" is described as a "loss of head" in the system, and can be calculated from graphs and charts provided by the pump and piping manufacturers. If a control valve is to be installed in the piping system for flow control, it is good practice to generate a series of system head curves for various valve positions, i.e. 25% open, 50% open, 75% open, and 100% open. The intersections of the various control valve position curves and the pump curve will allow the designer to predict the effect of valve position on flow and if the desired flow range will be achieved.

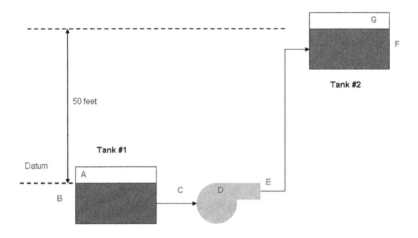

Figure 1.12-System head curve example

Refer to Figure 1.12 as we go through a simple system head calculation. Here we have a pump taking suction from Tank #1 ($P_{upsteam}$ = 0 psig) and moving to Tank #2 ($P_{downstream}$ = 0 psig) that has a liquid level that is 50 feet higher than the level in Tank #1. If we assume that the piping losses total 25 feet at 500 gpm, let's determine the system head requirements.

$$H_{system} = (P_{ds} - P_{us}) + H_{static} + L = (0-0) + 50 + 25 = 75 \text{ feet}$$

Keep in mind that the piping losses here only apply to a flow of 500 gpm. In reality the piping losses are a function of the flow rate, and therefore vary with the flowrate.

Let's go through another system head example where pressures must be considered. In Figure 1.13, we have a pump taking suction from an open tank on the left and discharging 200 gpm into a process vessel operating at 15 psig on the right.

$$H_{system} = (P_{ds}-P_{us}) + H_{static} + L$$

First, we will deal with the pressure terms. $P_{ds}-P_{us}$ = 15 psig. If the specific gravity of the liquid is 1.0, then this pressure term is equal to 15 x 2.31 = 34.65 feet. Next, we will assume the piping losses (L) are a constant 30 feet. In this example, we have two different liquid level conditions, A and B. At condition A, Hstatic = 30-6=24 feet, and at condition B, Hstatic= 38-2=36 feet. Therefore we have two different system heads for the two conditions:

Condition A System Head = 34.65+30+24 = 88.65 feet

Condition B System Head= 34.65+30+36 = 100.65 feet

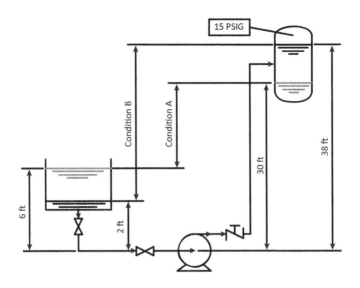

Fig. 1.13 – Pump discharging into a pressurized system

Based on this analysis, we would need a pump rated for at least 100.65 feet of head at 200 gpm. A control valve would be required to drop the pressure whenever condition A occurs.

The calculation of the fluid friction losses is beyond the scope of this book. However, a good "rule of thumb" to keep in mind when reviewing the calculation of piping losses is that piping losses are typically a function of flow squared, i.e. Losses=K x Q^2, where K is a constant. (Note: The reader will find Appendix A: How to Calculate Pipe Friction Head Losses for Newtonian Fluids a useful reference for determining piping friction losses. In addition, Hydraulic Institute Manuals and the Pump Handbook are excellent texts explaining how to perform detailed head loss calculations.) Rewriting the system curve equation we get:

$$H_{system} = (P_{ds}-P_{us}) + H_{static} + K \times Q^2$$

Since the terms $(P_{ds}-P_{us}) + H_{static}$ are constants once a system is defined, we can simplify the system term equation even further:

$$H_{system} = K_1 + K_2 \times Q^2, \text{ where } K_1 \text{ and } K_2 \text{ are constants.}$$

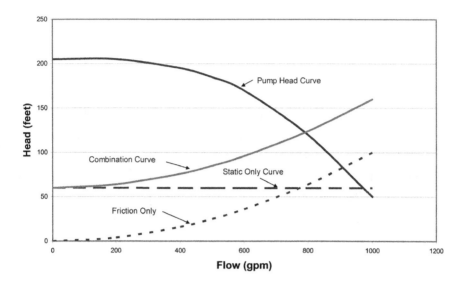

Figure 1.14-System Curves

Figure 1.14 depicts the three basic types of system curves: a) static only system curves, b) friction only system curves, and c) combination system curves. An example of a static only system is a pumping system where a pump is pumping from an open pit into an open pit at a higher elevation with negligible line losses. An example of a friction only system is a pump with significant piping losses that is taking suction from an open pit and discharging into an open pit that are at the same elevations. A combination system is a pump system with both significant frictional losses and a significant difference in elevation or pressure at the inlet and outlet piping. Combination curves tend to be defined by the simplified system curve equation:

$$H_{system} = K_1 + K_2 \times Q^2$$

By studying the simplified form of the system head equation, you can surmise that in general the system head equation has the form of a parabola, i.e. $y = A + B x^2$.

If we combine the system head equation and a centrifugal pump curve, you will get plots similar to what you see in Figure 1.15. Notice that in Figure 1.14 that we have one centrifugal pump head-flow curve and three system curves. For simplification, all the system curves are friction only type curves. There is a small piping curve, ideal piping curve, and a large piping curve.

As we stated before, the intersection of the pump head-flow curve and the system head curve is the point where the pump will operate. If a "small piping system" is installed, the pump will operate in the low flow region of the pump curve, and if a "large piping system" is installed the pump will operate in the high flow region of the pump curve. However, if the "ideal piping system" is installed the pump will operate at an ideal flow region, which is near its best efficiency point.

Normally the system curve is defined by the anticipated process conditions and tends to be set in stone once the process design step of the project is completed. It is the designer's job to take the system curve information and attempt to match a pump to the piping system. However, to select the right pump, we must have confidence in the system head curve, which is why it is a good idea to understand how system head curves are created and how they are affected by changes in process conditions. Once you clearly understand the shape of your system curve, you can begin choosing the right centrifugal pump to match your hydraulic requirements.

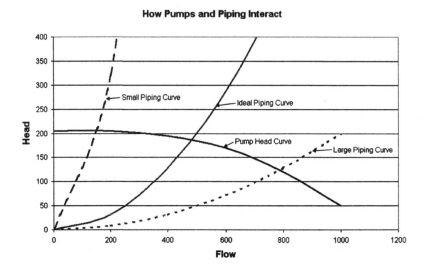

Figure 1.15-Pump and system interaction curve

Remember that given the right pump suction conditions, centrifugal pumps always pump somewhere on their curve. The shape of the pump curve will generally be determined by the specific speed number of the impeller. The manufacturer generates their curves at a specific RPM. If you are using a variable speed driver remember to adjust the curves to match your actual pump speed. You can use the pump affinity laws to approximate the expected performance at the actual pump speed. In addition to the specific speed parameter, the manufacturer can select impeller diameter, impeller width, pump rpm. You also have the options of operating pumps in series or parallel or selecting a multi-stage pump to meet your hydraulic needs.

The key to reliable pump operation is to ensure the pump and piping system are well matched to allow operation as close to the best efficiency point (BEP). To guarantee you have a well matched pump and piping system, you need to check and double check your system head calculations to confirm that all factors were considered and calculated properly. Your selection will only be as good as the system head equation and the manufacture's pump performance curve.

Chapter 2
Can I Use a Centrifugal Pump?

The first question you need to ask before selecting a centrifugal pump is: Will a centrifugal pump work in my application? There are many reasons to employ a centrifugal pump for your application, such as its simplicity, reliability, or ease of maintenance. If the right pump is selected and installed properly, you can expect a mean time of repair of between 3 to 7 years. However, there are several application factors that will immediately drop a centrifugal pump out of the running:

1. There is too much entrained gas or vapor in the fluid to be pumped

Virtually any type of centrifugal pump can handle some amount of entrained gas. The problem to be addressed is the tendency for the gas to accumulate in the pump suction inhibiting flow and head generation. If gas continues to accumulate, the pump may lose prime. Fig. 2.1 shows how the performance of a standard end suction pump is affected by various amounts of air. With a minor performance correction, centrifugal pumps are reasonably efficient in handling up to approximately 2%.

Some manufacturer's state they can handle up to 4% entrained gas, but you need to be careful at these high vapor levels. In practice, as the percentage of gas exceeds 4% by volume, the performance of a conventional pump begins to degrade drastically until the pump becomes unstable, eventually losing prime.

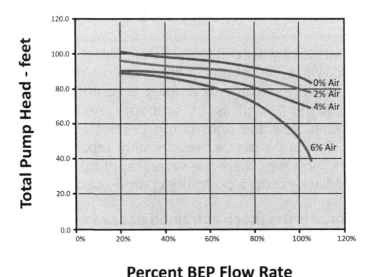

Percent BEP Flow Rate

Fig. 2.1 – Effect of vapor on centrifugal pump performance

2. **Solids in liquid are too large to pass through the pump impeller passages.**

The maximum allowable size of a solid particle in the pumpage is usually limited by the physical size of the minimum impeller opening. Usually the impeller exit channels represent the smallest openings. Strainers with openings that are smaller than the smallest impeller flow area can be installed in the suction piping to prevent plugging. However, if a preponderance of solids are present in the pumpage, the strainer will plug frequently and cause a maintenance nightmare.

In the municipal waste management applications, solids handling pumps are designed with the capability of handling specific spherical diameters. A few models of such pumps will have an open impeller while many will be designed with closed impellers having the necessary clearance between the vanes of the impeller and also between the shrouds.

> Tip: As a rule of thumb, assume a pump can pass a spherical solid that is one half of the smallest impeller opening. Consult the manufacturer if you expect a continuous stream of solids.

Although most designers tend to worry about the size of solid particles in the liquid being pumped, they sometimes overlook the major problem in trash applications that tend to contain stringy material. This material will invade the eye of the impeller, wrap itself around the shaft nut and eventually clog the pump.

Note: Slurries, which are a mixture of solid particles entrained in a liquid, can be handled by some centrifugal pump designs. If you need a pump to handle a slurry, make sure your supplier knows your industry and can provide a proven design that is engineered for your particular application. In slurry applications there's no standard design that can fit all applications.

If you anticipate a large volume of solids in the fluids stream, you need to select a seal or sealing system that can handle the solids content. Depending on the amount of solids, several sealing option are available to protect the seal or seals. You can either install a seal that resists fouling or you can install a flush plan that will isolate the seal from plugging.

3. The fluid viscosity greater than 3000 cSt

Viscosity negatively impacts pump performance by causing an increased resistance to flow throughout the pump. We are all familiar with how well honey adheres to the surface of a spoon. It falls off the spoon in a long strand, resisting separation from either itself or the spoon. The action separating the fluid stream from itself and the spoon surface is fluid shear. As gravity acts on the honey, it continuously shears the honey, causing it to slowly fall off the spoon. Similarly, the centrifugal action of a pump impeller generates shearing forces acting on the liquid as it moves through the pump. Viscosity increases the resistance to shearing forces generated by the impeller action and all the pump's internal surfaces. The result is a loss of capacity and head and increased power consumption.

Most pumped fluids are classified as Newtonian fluids where the resistance to shear is directly proportional to the shearing force applied. The amount of performance loss from viscous effects is dependent on pump geometry, speed, and the viscosity of the fluid pumped. 3000 cSt is often given as the practical limit of centrifugal pump under viscous conditions, but the actual limit is geometry dependent and the pump manufacturer should be consulted. (For comparison of various fluid viscosities, refer to Table 2.1 below. Notice that honey has a viscosity of 73.6 cSt at 100F, which is over 73 times more viscous than water at 68 F.) High flow pumps move very large flows relative to impeller and casing surface areas and thereby are the least effected by viscosity. Low flow impellers have narrow fluid passages that have large surface areas resulting in high frictional forces relative to the amount of fluid pumped.

Table 2.1- Comparison of Various Fluid Viscosities

Liquid	Temperature		Kinematic Viscosity	
	(°F)	(°C)	CentiStokes (cSt)	Seconds Saybolt Universal (SSU)
Castor oil	100	37.8	259-325	1200-1500
	130	54.4	98-130	450-600
Cod oil (fish oil)	100	37.8	32.1	150
	130	54.4	19.4	95
Corn oil	130	54.4	28.7	135
	212	100	8.6	54
Cotton seed oil	100	37.8	37.9	176
	130	54.4	20.6	100
Crude oil 48° API	60	15.6	3.8	39
	130	54.4	1.6	31.8
Crude oil 40° API	60	15.6	9.7	55.7
	130	54.4	3.5	38

Crude oil 35.6° API	60	15.6	17.8	88.4
	130	54.4	4.9	42.3
Crude oil 32.6° API	60	15.6	23.2	110
	130	54.4	7.1	46.8
Glycerine 100%	68.6	20.3	648	2950
	100	37.8	176	813
Honey	100	37.8	73.6	349
Kerosene	68	20	2.71	35
Jet Fuel	-30	-34.4	7.9	52
Linseed oil	100	37.8	30.5	143
	130	54.4	18.94	93
Tar, pine	100	37.8	559	200-300
	132	55.6	108.2	55-60
Water, distilled	68	20	1.0038	31

Figure 2.2 illustrates how viscosity affects centrifugal pump performance. The solid lines represent pump performance without viscosity effects and the dotted lines represent pump performance due to a 1000 SSU fluid. Notice how the developed head falls below the theoretical head curve and that efficiencies fall far below those expected for inviscid fluids. As a consequence of the lower expected efficiencies, the predicted power levels rise above the manufacturer's standard horsepower curve.

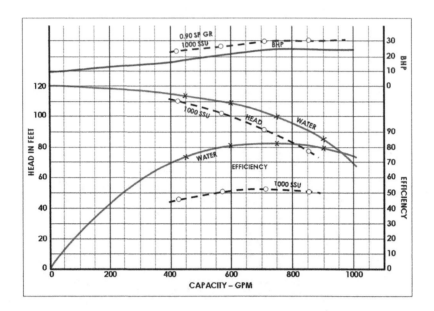

Fig. 2.2-Effect of Viscosity on Performance

Table 2.2 is a summary of the effects of viscosity on three different centrifugal pumps. Each pump is represented with two columns (expected efficiency and efficiency loss due to viscosity). As an example, let's look at a 250 gpm, 200 foot pump that is pumping a 100 cSt viscosity liquid. Under these conditions, the efficiency drops from 75% down to 52%. Now let's see what happens when this same pump has to pump a 1000 cSt liquid. The efficiency drops from 75% down to 17%.

You can see several trends by studying this table: First, you see that efficiencies drop off as viscosities increase. You also notice the performance of smaller pumps tend to be affected more significantly by higher viscosity liquids. Finally you can see that there is a point when efficiencies start to drop off dramatically.

Table 2.2: Comparison of expected centrifugal pump efficiency various viscosities. Assumptions: SG= 0.9 and Baseline Eff=75%

Viscosity (Cst)	100 gpm @150 ft: expected efficiency	100 gpm @150 ft: efficiency loss	250 gpm @ 200 ft: expected efficiency	250 gpm @ 200 ft: efficiency loss	1000 gpm @ 325 ft: expected efficiency	1000 gpm @ 325 ft: efficiency loss
10	70.0%	6.7%	69.0%	8.0%	72.0%	4.0%
50	56.0%	25.3%	59.0%	21.3%	65.0%	13.3%
100	46.0%	38.7%	52.0%	30.7%	60.0%	20.0%
250	29.0%	61.3%	40.0%	46.7%	51.0%	32.0%
500	15.0%	80.0%	30.0%	60.0%	43.0%	42.7%
1000	7.0%	90.7%	17.0%	77.3%	32.0%	57.3%
2000	4.0%	94.7%	7.0%	90.7%	18.0%	76.0%
3000	0.0%	100.0%	8.0%	89.3%	11.0%	85.3%

As we have shown here, higher liquid viscosities can significantly affect pump performance, which requires pumps and drivers to be oversized. There is a point when a centrifugal pump may not make economic sense. Carefully evaluate the total life cycle cost when considering using a centrifugal to pump a viscous liquid.

Tip: While many centrifugal pump manufacturers say they can handle fluids with viscosities 1,000 cSt and higher, PD pumps tend to be a better choice above these viscosity levels when considering the high energy costs resulting from the loss of efficiency.

4. The pressure at the suction flange is too close to the fluid's vapor pressure

For a centrifugal pump to operate properly, it must be provided with a supply of fluid above its vapor pressure. As the fluid travels from the suction vessel towards the pump's suction, its pressure will drop due to frictional losses. If the fluid in the suction vessel is near its vapor pressure or bubble point, there is a risk that the fluid may vaporize as it travels down the piping. To ensure the fluid does not vaporize, you must maintain a margin between the actual pressure at the pump's suction and the fluid's vapor pressure at the pumping temperature. This margin is called the net positive suction head (NPSH). For a centrifugal pump to function reliability, the net positive suction head available (NPSHa) at the pump suction flange must be greater that the net positive suction head required (NPSHr) by the pump.

Net Positive Suction Head - NPSH

The net positive suction head (NPSH) available (NPSHa) is a function of the suction piping system in which the pump operates. Simply put, it is the excess pressure of the liquid in feet absolute over its vapor pressure as it arrives at the pump suction. By definition, net positive suction head (NPSH) available can be expressed as the difference between the suction head at the pump suction flange and the liquids vapor head and expressed as follows:

$$NPSH = h_s - h_v$$

To properly calculate NPSHa you must be able to understand and evaluate your particular suction piping system. Suction piping systems come in different configurations depending on the specific needs of the process. There are four basic types of suction systems:

1. Flooded suction from an atmospheric tank (see Figure 2.3)
2. Flooded suction from a pressurized vessel
3. Suction lift from an open pit (see Figure 2.4)
4. Suction lift from a pressurized system

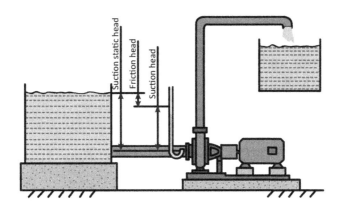

Fig. 2.3 - Pump with a flooded suction

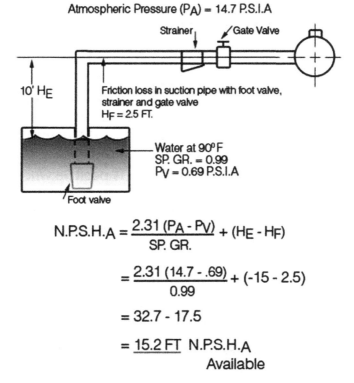

$$N.P.S.H._A = \frac{2.31 \, (P_A - P_V)}{SP. \, GR.} + (H_E - H_F)$$

$$= \frac{2.31 \, (14.7 - .69)}{0.99} + (-15 - 2.5)$$

$$= 32.7 - 17.5$$

$$= \underline{15.2 \, FT} \quad N.P.S.H._A$$
$$\text{Available}$$

Fig. 2.4 - Pump with a suction lift

The net positive suction head available is usually determined during design and construction phase of a project, but can also be determined experimentally from the actual physical system. The precise calculation of the NPSHa value is highly dependent on the type of suction system being analyzed and the physical arrangement of the system.

The equation for the NPSHa is:

$$NPSH_a = H_A \pm H_Z - H_F + H_V - H_{VP}$$

The variables in the NPSHa equation are defined below. The notes below are provided to help the reader better understand how each term is used in the equation:

Term	Definition	Notes
H_A	The absolute pressure on the surface of the supply tank	1. Typically atmospheric pressure is assumed. 2. Don't forget the altitude effects 3. Remember that even vacuum vessels have a positive absolute pressure
H_Z	The vertical distance from the surface of the liquid in the supply tank and the pump centerline	1. Is positive when the liquid level is above the pump centerline and negative if the liquid level is below the pump centerline. 2. Always use the lowest liquid level expected in the tank or vessel.

H_F	Friction losses in the suction piping	1. Piping and fittings act as flow restrictions that reduce the liquids pressure in the suction piping.
H_V	Velocity head	1. This is typically a small value that can be ignored.
H_{VP}	Absolute vapor pressure of the liquid at the pumping temperature	1. Always subtract this value in the NPSHa equation. 2. Remember that when the operating temperature goes up the vapor rises. 3. Always use the highest expected operating temperature to determine the maximum possible vapor temperature.

We will go through a few NPSHa calculation examples to illustrate how the NPSHa equation is modified to handle the various suction arrangements.

Available NPSH$_a$ - Pump is below a pressurized tank

Here, a pump is taking suction out of an overhead, pressurized tank with a liquid level that is 20 feet above the pump centerline. We assume the specific gravity is 0.9.

We start with the NPSHa equation and enter the appropriate inputs as shown in Table 2.3 below.

$$NPSH_a = H_A \pm H_Z - H_F + H_V - H_{VP}$$

Here are the inputs:

Table 2.3-NPSHa example #1

Term	Definition	Inputs
H_A	The absolute pressure on the surface of the supply tank	100 psi=2.31*100/0.9=256.7 feet
H_Z	The vertical distance from the surface of the liquid in the supply tank and the pump centerline	+20 feet
H_F	Friction losses in the suction piping	-15 feet
H_V	Velocity head	0
H_{VP}	Absolute vapor pressure of the liquid at the pumping temperature	95 psi=2.31*120/.9=243.8 feet

Inputting all the values we get:

$$NPSH_a = H_A \pm H_Z - H_F + H_V - H_{VP} = 256.7+20-15-243.8=17.9 \text{ feet}$$

Available NPSH$_a$ - Pump is below an atmospheric tank (see Figure 2.3)

A pump is taking suction out of an atmospheric tank with a liquid level that is 20 feet above the pump centerline. Assume the specific gravity is 0.9.

Again, we start with the NPSHa equation and enter the appropriate inputs as shown in Table 2.4 below

$$NPSH_a = H_A \pm H_Z - H_F + H_V - H_{VP}$$

Here are the inputs:

Table 2.4-NPSHa example #2

Term	Definition	Inputs
H_A	The absolute pressure on the surface of the supply tank	14.7 psi=2.31*100/0.9=37.73 feet
H_Z	The vertical distance from the surface of the liquid in the supply tank and the pump centerline	+20 feet
H_F	Friction losses in the suction piping	-15 feet
H_V	Velocity head	0
H_{VP}	Absolute vapor pressure of the liquid at the pumping temperature	0.5 psi=2.31*120/.9=1.28 feet

Inputting all the values we get:

$$NPSH_a = H_A \pm H_Z - H_F + H_V - H_{VP} = 37.73+20-15-1.28=41.45 \text{ feet}$$

Available NPSH$_a$ - Pump is above an open (atmospheric) pit (see Figure 2.4)

A pump is taking suction out of an open (atmospheric) pit with a liquid level that is 10 feet below the pump centerline. Assume the specific gravity is 0.9.

We start with the NPSHa equation and enter the appropriate inputs as shown in Table 2.5 below.

$$NPSH_a = H_A \pm H_Z - H_F + H_V - H_{VP}$$

Here are the inputs:

Table 2.5-NPSHa example #3

Term	Definition	Inputs
H_A	The absolute pressure on the surface of the supply tank	14.7 psi=2.31*100/0.9=37.73 feet
H_Z	The vertical distance from the surface of the liquid in the supply tank and the pump centerline	-10 feet
H_F	Friction losses in the suction piping	-15 feet
H_V	Velocity head	0
H_{VP}	Absolute vapor pressure of the liquid at the pumping temperature	0.5 psi=2.31*120/.9=1.28 feet

Inputting all the values we get:

$$NPSH_a = H_A \pm H_Z - H_F + H_V - H_{VP} = 37.73\text{-}10\text{-}15\text{-}1.28 = 11.45 \text{ feet}$$

Required NPSH - NPSH$_r$ or NPSHR

NPSHa is associated with the suction piping system. The pump is assigned with critical parameter called NPSHr, which stands for the net positive suction head required by the pump. While these terms look very similar, they are really vastly different in physical meaning. NPSHr is the liquid head above the vapor pressure required at the pump suction flange to prevent cavitation and ensure safe and reliable operation.

The required NPSH$_r$ for a particular pump is in general determined experimentally by the pump manufacturer and a part of the documentation of the pump. Most pump performance curves have an NPSHr curve plotted over the design flow range of the pump (see Figure 2.5).

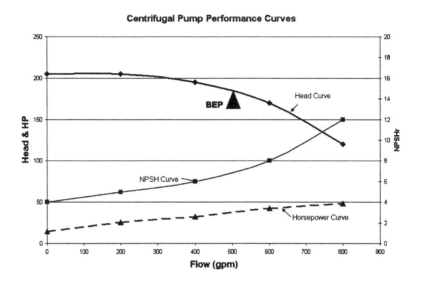

Figure 2.5 - Pump curve with NPSHr curve

The available $NPSH_a$ of the system should always exceeded the required $NPSH_r$ of the pump in order to avoid vaporization and cavitation of the impellers eye, i.e. NPSHa > NPSHr. The available $NPSH_a$ should in general be significantly higher than the required $NPSH_r$ so that the head losses in the suction pipe and local velocity accelerations in the pump casing do not lead to flashing in the pump impeller. It is a good practice to add a NPSH margin to account for process upsets that may erode the NPSH available.

Note that the required $NPSH_r$ increase with the square of the flow which means that cavitation is more likely to occur at higher flowrates.

Pumps with double-suction impellers have lower $NPSH_r$ requirements than pumps with single-suction impellers. Although pumps with double-suction impellers are considered hydraulically balanced, they tend to be susceptible to uneven and unstable flow, particularly if the suction piping is not carefully designed.

| Tip: Pumping Hydrocarbons |

Be aware that the NPSH specification provided by the manufacturer in general is based on testing with cold water. For hydrocarbons, manufacture's test stand values must be lowered to account for the vapor release properties of complex organic liquids.

Note that the head developed by a pump is independent of the liquid, and that the performance curves for water from the manufacturer can be used for Newtonian liquids like gasoline, diesel or similar. Be aware that required power depends on liquid density and must be adjusted.

Ways to increase the margin between the NPSHa and the NPSHr

The reader needs to understand that there are practical methods available to improve the margin between the system's NPSHa and the pump's NPSH requirement to avoid the possibility of cavitation. You can either increase the NPSH available or decrease the NPSH required to improve the NPSH margin. We will discuss three "knobs" that a pump user can turn to affect the pump's NPSHr and four "knobs" to affect the NPSH available.

The three "knobs" the pumping system designer can turn to decrease the NPSHr requirements are:

1. Reduce the impeller speed to decrease the NPSHr
2. Increase the diameter of the impeller suction eye to decrease the NPSHr. (Note that there is a practical limit to increasing the impeller diameter. At this point, we will limit the suction eye diameter by limiting the impeller

design parameter Nss to a maximum value of 11,000. This will be discussed in more detail later in the book.)

3. Use a double suction impeller to decrease the NPSHr. Installing a double suction impeller is equivalent to increasing the impeller suction eye area.

The four "knobs" that the pumping system designer can use to increase the system's NPSHa are:

1. Elevate the liquid level in the suction source
2. Increase the size of the suction piping to decrease the overall pressure drop
3. Add cooling to sub-cool the liquid entering the suction vessel to reduce its vapor pressure.
4. If possible, increase the pressure over the liquid level in the suction vessel to increase the NPSHa.

All of these NPSHa improvement options require some amount of additional capital investment.

Let's examine how impeller geometry and speed can affect the NPSHa. Figures 2.6 and 2.7 summarize the minimum net positive suction head that is required (NPSHr) for a given flow and design condition. This figure contains four different curves: NPSHr versus flow for an 1800 rpm, single suction pump, 3600 rpm, single suction pump, an 1800 rpm, double suction pump, and a 3600 rpm, double suction pump.

For example, a 6000 gpm with a single suction eye, operating at 3600 rpm will require a minimum of about 75 feet of NPSHr. Dropping the speed down to 1800 rpm will allow you to drop the NPSH requirement down to about 28 feet with a single suction impeller. (Note that this NPSHr chart is based on an accepted design practice of keeping the suction specific speed below a recommended value of below 11,000. Suction specific speed will be discussed later in the book.)

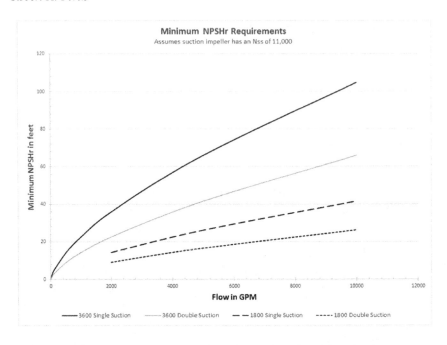

Figure 2.6- NPSHr versus flow for various impeller designs

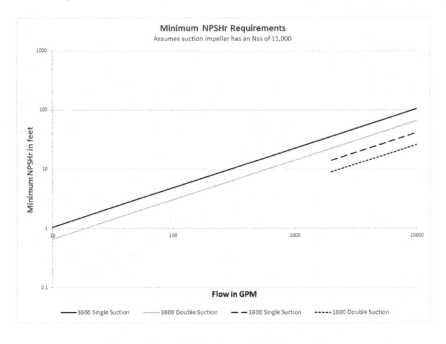

Figure 2.7- NPSHr versus flow for various
impeller designs in log-log format.

These charts show that a lower NPSHr value can be obtained by either lowering the speed or doubling the impeller eye area by using a double suction impeller.

5. The flow-head requirements don't fall within the normal design coverage of centrifugal pumps

There are many pump designs available for any given task. Pump designers need a way to compare the performance and efficiency of their designs across a large range of pump model and types. As a means of comparison, centrifugal pumps have been tested and evaluated using a dimensionless number called the specific speed (N_s). (Ns will be explained in more detail in the next chapter.) The efficiency of pumps with the same specific speed can be compared providing the user or the designer a starting point for comparison or as a benchmark for improving the design and increasing the efficiency. The equation for the pump specific speed (shown below) requires you to know H, the pump total head; N, the rotational speed of the impeller; and Q the flow rate.

$$N_s = \frac{n(rpm) \times \sqrt{Q(USgpm)}}{H(ft\ fluid)^{0.75}}$$

From a hydraulic design point of view, pumps traditionally fall into 3 categories; radial flow, mixed flow and axial flow (see Figure 2.8). There is a continuous spectrum of designs from the radial flow impeller, which develops pressure principally from the action of centrifugal force, to the axial flow impeller, which develops most of its head by the propelling or lifting action of the vanes on the liquid.

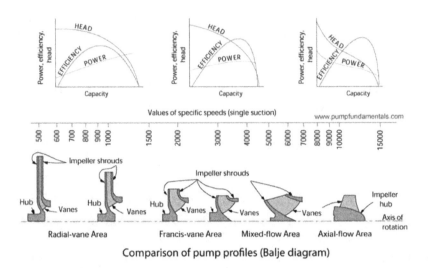

Comparison of pump profiles (Balje diagram)

Figure 2.8 – How impeller geometry
is related to specific speed

Specific speed has also been used as a design criterion for evaluating the efficiency of standard volute pumps (see Figure 2.9). You will notice that larger pumps are inherently more efficient than smaller pumps and that efficiency drops rapidly at specific speeds of 1000 or less.

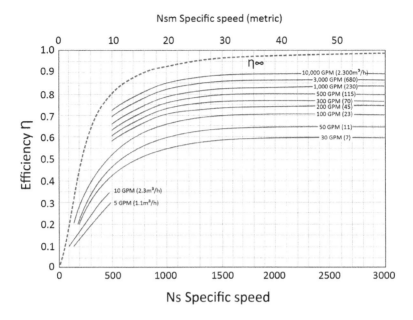

Fig. 2.9 – How efficiency varies with specific speed and flow
If the Ns value ultimately falls below 500, then you probably won't find a centrifugal pump for your application. It is possible to increase the Ns value by adding pumping stages or by increasing the pump speed; however these options may not always be available or make economic sense.

Summary

There are five factors that have the ability to eliminate centrifugal pumps from consideration. They are:

1. There is too much entrained gas or vapor in the fluid to be pumped
2. Solids in liquid are too large to pass through the pump impeller passages.
3. The fluid viscosity greater than 3000 cSt

4. The pressure at the suction flange is too close to the fluid's vapor pressure

5. The flow-head requirements don't fall within the normal design coverage of centrifugal pumps

If none of these factors have knocked your application out of the running for a centrifugal pump, then it's time to fine tune your selection.

Chapter 3
How Good Pumps Become Bad Actors

I have seen, and I am sure you have too, countless examples of installed pumps that are poor hydraulic fits. These selection mistakes are called off-design pump selections, bad actors, misapplied pumps, etc. We are all familiar with these pumps because they live in our repair shops and require inordinate amounts of time and care.

After much anguish and frustration, someone always asks the inevitable question: How did this pump ever make it into our plant? There are many reasons why these dubious selections make it into the field, including:

1. The pump supplier did not have a good hydraulic fit to offer, so he offered what he had available.
2. The user badly missed the actual flow requirements, so the pump manufacturer cannot be blamed.
3. More costly options were not considered, such as multistage or lower speed designs, as a means to keep initial capital costs down.

For example: Due to a low $NPSH_a$ specification, a pump supplier chose to bid a 3600-rpm pump that must operate well below its best efficiency point in order to keep the $NSPH_r$ low, instead of considering an 1800-rpm option. This selection "trick" places the pump between a rock and a hard place—dooming the pump to operate at low flow conditions forever! At lower flows, the pump will become unstable. At higher flows, the pump will cavitate.

Remember that there are usually at least four parties involved in the pump selection process: the process engineer, the project engineer, the rotating equipment engineer, and the pump supplier. Each party has a different point of view and different objectives.

Pump Selection Philosophies

In no particular order, I will briefly outline how each team member views pump selection:

The *process engineer* always wants to ensure the pump meets his or her process requirements, so he or she will typically add a safety margin to the flow and $NPSH_a$ specifications. These safety margins usually move the actual operating point away from the pump's best efficiency point and increase the pump's required suction specific speed. There is also some concern about energy efficiency, but this is usually secondary to ensuring the process requirements are satisfied.

The *pump supplier* wants to sell you the lowest cost pump possible so that he or she is the lowest bidder. If cost was no object, he would want to sell you the most expensive pump he has. But this is not the case. Due to extreme bidding competition, the pump supplier's goal is to bid the cheapest pump that meets your technical requirements and the manufacturer's selection guidelines.

The *project engineer* wants the least expensive pump that meets his design requirements. This means his or her objectives closely match those of the pump suppliers. However, by soliciting competitive bids from several pump suppliers, the project engineer guarantees he or she will obtain the lowest cost pump for the provided technical specifications.

The *rotating equipment engineer* wants his or her pump to be as reliable as possible. He does not want to spend his nights and weekends fixing or worrying about this pump. Experience has taught him or her to:

1. Always select a pump with a healthy NPSH margin to protect the pump during process upsets.
2. Always select a pump that fits the process requirements well. In other words, only select a pump that will operate in the middle of its pump curve, where the

efficiency is the highest and flows tend to be more stable. Hopefully, select a pump that will always operate between 70 percent to 80 percent of their best efficiency point (BEP).

3. Always select the slowest viable pump speed to ensure long-term mechanical reliability, i.e. operation at 1800-rpm is always better than operation at 3600-rpm. However, this is always a difficult sell. Slower always means more expensive, since slower pumps are larger, hence requiring more metal to construct, leading to the great chasm of pump selection. The project people like 3600-rpm pumps because of their lower cost, and the rotating equipment engineers like 1800-rpm pumps because of their improved reliability.

4. Always select a pump that satisfies your relevant design spec, i.e. API 610, ASME B73.1, ASME B73.2, etc.

5. To ensure your pump achieves the highest level of reliability and safety, select mechanical seals that conform to API 682 (ISO 21049). This specification is considered the leading document for mechanical seals in petrochemical, chemical, and pipeline services worldwide. (Refer to Appendix B: Mechanical Seal Selection Design Primer to learn more about mechanical seals.)

All these requirements come with a price, one that the other selection team members may not think is justified. This is what makes the pump selection process fun!

What situation does this put the rotating equipment engineer? He is placed into a debate of one against three, with the rotating equipment engineer having the least control over the final outcome. He has no money. He only has decision rights if empowered by his or her organization. This should not be considered a power struggle, but a collaborative selection process with the long-term benefits of the organization in mind.

Pump selection should only be considered a win-win if it represents the lowest cost of ownership over the project's life.

What does this mean? It means if you add up the cost of the pump, installation costs, and energy costs over its lifetime, repair costs, and cost of unreliability for all pumps considered, the pump selected should be the least expensive to own. If it is not, then the entire team failed to do its job.

But there is a catch in this reasoning. You only get to consider the pumps that are quoted. So you ask: How can I consider pumps that are not quoted? You cannot, but you *can* influence the types of pumps that are quoted - and I will show you how.

Let's look at the two equations that can be used to analyze your pump design options, the equations for specific speed and the equation for suction specific speed. First I will cover specific speed.

Specific Speed

Specific speed (Ns) is a term used to describe the geometry (shape) of a pump impeller. Those responsible for the selection of the proper pump can use this Specific Speed information to:

- Select the shape of the pump curve.
- Determine the efficiency of the pump.
- Select the lowest cost pump for their application.

Specific speed is defined as "the speed of an ideal pump geometrically similar to the actual pump, which when running at this speed will raise a unit of volume, in a unit of time through a unit of head".

The performance of a centrifugal pump is expressed in terms of pump speed, total head, and required flow. This information is obtained from the pump manufacturer's published curves. Specific speed is calculated from the following formula, using data from these curves at the pump's best efficiency point (BEP):

$$Ns = (n \times \sqrt{Q}) / (H / stg) \wedge 0.75$$

Here is what the variables in the Ns equation mean:

n = The speed of the pump in revolutions per minute (rpm.)

Q = The flow rate in gallons per minute (for either single or double suction impellers)

H = The total dynamic head in feet. H can be calculated by knowing the pump's differential pressure and the specific gravity (SG) of the liquid. H=(Pd-Ps)*2.31/SG.

Stg=The number of stages

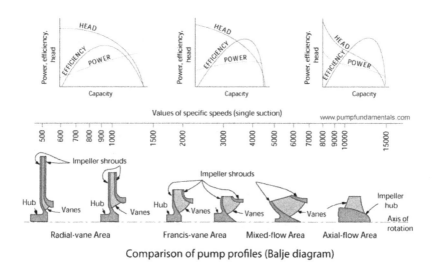

Comparison of pump profiles (Balje diagram)

Figure 3.1

Pumps are traditionally divided into three types: radial flow, mixed flow, and axial flow. (Refer to the chart above.) When you look at the chart you can see there is a gradual change from the radial flow impeller, which develops pressure principally by the action of centrifugal force, all the way to the axial flow impeller, which develops most of its head by the propelling or lifting action of the vanes on the liquid.

The most commonly used impellers fall in the specific speed range from approximately 1000 to 6000. Impellers can be either double suction impeller or single suction designs.

$$Ns = (N \times \sqrt{Q}) / (H/stg)^{0.75}$$

Let's analyze Figure 3.1 in some detail in order to determine how Ns affects head, efficiency, and power consumption. We can see that:

> ➢ The steepness of the head/ capacity curve increases as specific speed increases.
> ➢ At low specific speed, power consumption is lowest at shut off and rises as flow increases, which means that the driver has a tendency to be over loaded at the higher flow rates. This characteristic of radial flow pumps should be evaluated at the time of purchase.
> ➢ At medium specific speed the power curve peaks at approximately the best efficiency point and then falls off at higher flows. This non-overloading feature enables the pump to operate safely over most of the fluid range with a driver rated for the BEP flow conditions.
> ➢ High specific speed pumps have a falling power curve with the maximum power demand occurring at low flow conditions. For this reason, these type of pumps should never be started with the discharge valve shut. If throttling is required for the application, the driver should be sized for minimum flow conditions.
> ➢ Rules of thumb: a) lower specific speeds produce flatter curves, while higher specific speed pumps produce steeper ones, b) lower specific speeds have continuously rising horsepower curves and high specific speed pumps have continuously falling horsepower curves.

Keep in mind that efficiency and power consumption were calculated at the best efficiency point (BEP). In practice most pumps operate in a throttled condition because the pump was oversized at the time it was purchased. Lower specific speed pumps may have lower efficiency at their BEP, but at the same time will have lower power consumption at reduced flow than many of the higher specific speed designs.

The result is that it might prove to be more economical to select a lower specific speed design if the pump has to operate over a broad range of capacity.

Tip:

Hydraulic efficiency peaks at specific speeds (N_s) between 2000 and 3000 (in US units) and drops dramatically below 500 (in US units). We can glean several design ideas from this rule of thumb. First, if efficiency is important to you, try to select pumps with specific speeds in the 2000 to 3000 (in US units) range. Second, if your initial calculation of Ns falls below 500 (in US units), you either need to consider more impeller stages, increase your speed, or consider another type of pump, such as plunger, screw, or diaphragm pump.

Suction Specific Speed

Suction Specific Speed is another dimensionless parameter used when characterizing impellers and selecting centrifugal pumps. While specific speed (Ns) is mostly related to the discharge side of the pump, the suction specific speed deals primarily with its suction (inlet) side. The head (H) term in the denominator of the defining formula for the Ns is substituted by the NPSHR:

$$Nss = \left(n \times \sqrt{Q / (1 or 2)} \right) / (NPSHr)^0.75$$

Here is what the variables in the Ns equation mean:

n = The speed of the pump in revolutions per minute (rpm.)

Q/(1 or 2) = The flow rate in gallons per minute (Divide by either 1 for a single suction impeller or 2 for a double suction impeller)

NPSHr = The net positive suction head required in feet

Values of Nss typically vary from about 6,000 to 15,000, and sometimes even higher for the specialized designs.

A user would prefer to provide as low NPSHa as possible, since a lower NPSHa usually corresponds to a lower system cost. For example:

1. A higher level of liquid in the basin of the cooling water pumps requires taller basin walls
2. A deeper excavation to lower a pump centerline below the liquid level.

However, a pump manufacturer wants to have ample NPSHa, with a significant margin above the pump NPSHr (net positive suction head required), to avoid cavitation, damage, and similar problems. A wider NPSH margin (M) is achieved either by increasing the NPSHa, or decreasing the NPSHa, since

$$M = NPSHA - NPSHR$$

An end user may think that a lower NPSHr design is preferable because it usually translates to lower equipment and construction costs and a pump manufacturer might offer a lower NPSHr design because they believe this is what customers are looking for. However, since a lower NPSHr design leads to a higher value of Nss (according to the equation above), the highest Nss design might seem desirable, but in reality, there are inherent problems related to high Nss designs.

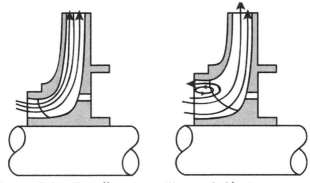

Figure 3.2a: Smaller eye   Figure 3.2b: Larger eye

Let's look at a low Nss impeller (Figure 3.2a) and a high Nss impeller (Figure 3.2b) side by side. Both impeller designs deliver the same head and capacity. Notice that the impeller in Figure 3.2b has a larger eye than impeller in Figure 3.2a and will therefore require a lower NPSHr due to the lower pressure drop in the impeller eye. The lower NPSHr comes at a cost however, as shown by the unstable suction flow in Figure 3.2b. By opening the impeller eye, the designer has increased the risk of flow recirculation occurring at the vanes (Figure 3.2b). Flow recirculation can result in cavitation, which can lead to vibration, noise and metal loss.

By attempting to provide a low NPSHr impeller to cope with a low NPSHa system, a potentially more serious problem can be created. If a pump operates below its BEP, the velocity profile changes, and no longer can maintained its uniformity and order. Fluid particles then begin to separate from the path of the sharpest curvature (which is the impeller shroud area), and the resulting mixing and wakes produce a turbulent, disorderly flow regime, which makes matters difficult from the NPSHR standpoint. The further left of the Best Efficiency Point (BEP) the pump is operated, the more exaggerated recirculation becomes.

We can conclude by starting that a larger impeller eye does decrease the NPSHR at the BEP point, but causes flow separation problems at the off-peak low-flow conditions. In other words, a high Suction Specific Speed (N$_{ss}$) design is better only if a pump does not operate significantly below its BEP point.

In the real world, there are few cases where centrifugal pumps operate strictly at its BEP flow. The flow demands at the plants change constantly, requiring control valves to continuously modulate pump flows. There are numerous examples of pumps with high N$_{ss}$ designs that are known reliability problems because of such frequent operation in undesirable low flow regions.

Actual plant studies have shown that above Nss of 11,000 pump mechanical reliability begins to suffer – exponentially. Realizing this fact, around mid-80s, users started to limit the value of the Nss, and a Hydraulic Institute uses Nss = 8500 as a typical guiding value.

NPSH Margin

To operate in a cavitation free manner, pumps need a margin of additional NPSH above the test stand values. It has been demonstrated that the amount of NPSH margin depends on the suction energy of the pump, which reflects energy available for cavitation damage. Suction energy is a function of the suction specific speed (Nss) of the pump.

Table 3.1 list the suggested NPSH margin ratio that should be maintained between the NPSHA and the NPSHR as recommended by the Hydraulic Institute.

TABLE 3.1
NPSH MARGIN RATIO GUIDELINES (NPSHA/NPSHR)

SUCTION ENERGY LEVEL	
Suction Energy Level	Margin
Low	1.1-1.3
High	1.3-2.0
Very high	2.0-2.5

Margin is defined as the ratio of NPSHA to NPSHR. The suction energy is a function of the momentum of the fluid approaching the impeller eye and is defined by the following equation:

Suction Energy = D_{eye} x n x Nss x Sp.Gr.

where:
D_{eye} = Impeller eye diameter (in)
n = Pump rotational speed (rpm)
Nss = Suction specific speed (see above)
Sp.Gr. = Specific gravity of the pumped fluid

Threshold Values for End Suction Pumps are defined as:
High suction energy 160 x 10^6
Very high suction energy 240 x 10^6

Here is example NPSH suction margin problem:

Let's say we have an impeller with a 6" eye, turning at 3600 rpm, with a specific gravity of 0.8 and a Nss of 10,500. The suction energy level is:

$$SE=6 \times 3600 \times 10,500 \times 0.8 = 181.4 \times 10^6$$

This SE value puts us in the "High" suction energy range, which means that an NPSH margin ratio between 1.3 and 2 is warranted. So, if an impeller has a 20 ft NPSH requirement, then you need at least 26 to 40 ft of NPSH to operate safely.

Tip: Here are a few guidelines to consider:

- Any value less than the high suction energy threshold is low suction energy
- Threshold values for split case/radial inlet pumps are 0.75 x end suction values
- Threshold values for vertical turbine pumps are 1.3x end suction values.
- For other pump rotational speeds, suction energy varies directly with the rpm ratio

---

It is generally good practice to add an additional 2-5 feet of NPSHA over the margin values to account for disparities between test data and actual site conditions.

The reader should be aware that cavitation is not the same as gas entrainment. Often times, especially with vertical pumps, a pump that is pulling air into the suction will be described as cavitating. Although the symptoms are similar the cure is not the same.

Remember that changes in process conditions such as increased flow or temperature, and physical changes to the suction side of the pump such as installation of upstream equipment, modification of pipe routes, or failure to maintain upstream equipment such as heat exchangers, may create cavitation problems. In order to keep capital costs low many system designs provide for adequate NPSH margins, but not much of a cushion beyond that. Therefore, any proposed changes to a system, or its operation, should be scrutinized carefully to be sure that adequate suction conditions are maintained.

NPSHr Corrections for Vertical Pumps

Vertical multistage centrifugal pumps, which are sometimes called vertical turbine (VT) and deep well pumps, (see Figure 1.9) are often used in sumps to ensure there is adequate submersion depth below the liquid level. Vertical turbine pumps can also be used inside a pressure containing can to increase the suction head available to the first stage impeller in order to increase the NPSHa and avoid cavitation. Keep in mind that the standard Nss equation shown here does not apply to VT pumps because the NPSHr term normally applies to pumps where the impeller centerline is at or close to the same elevation as the suction flange (see the pump shown in Figure 1.5)

$$Nss = \left(n \times \sqrt{Q/(1 \ or \ 2)} \right) / (NPSHr)^{\wedge} 0.75$$

If a VT pump is being considered, the elevation difference between the suction flange and the lowest impeller must be added to the NPSHr term in this equation to obtain a true Nss value for the first stage impeller.

Now that you have a better understanding of specific speed and suction specific speed, let's run through a few examples.

Table 3.2 shows the input and outputs from a spreadsheet I developed to simultaneously calculate Ns, Nss, NPSHr, NPSH ratio, and brake horsepower. (The reader can either reconstruct this spreadsheet calculator using the equations in this book or request a copy of my centrifugal pump analysis spreadsheet by writing me at rxperez1665@msn.com. Please use "Centrifugal Pump Analysis Spreadsheet" as the email subject.)

Table 3.2- Pump Design Analysis Table (3580 rpm Pump)

Inputs	Units	
RPM		3580
Flow @ BEP	gpm	500
Suction pressure	psi	25
Discharge pressure	psi	380
Specific gravity		0.95
Pump efficiency		0.64
NPSHa	ft	12
NPSH margin	ft	2
Number of stages		1
Number of suction eyes		1

Outputs	Units	
NPSHr	ft	10
NPSH ratio		1.2
Pump head	ft.	863.21
Head per stage		863.21
Ns		502.67

Nss		14235.35
Impeller suction eye diameter (estimate)	inches	6.00
Suction energy		2.90E+08
Brake horsepower	HP	161.78

The output from the example analysis shown in Table 3.2 tells you that a single stage pump, operating at 3580-rpm, with one suction eye, will require an impeller with an N_s of 503 (radial vane impeller design) and N_{ss} of 14,235. You quickly realize the N_{ss} here is much too high. Also notice that the NPSH ratio is 1.2, which is too low for suction energy of 290 x 10^6. According to the table below a suction energy of 290 x 10^6 requires an NPSH margin ratio of 2 to 2.5, which means the margin in this example falls well below the recommended. We can conclude that this example highlights several design issues: The Ns is on the borderline of being too low, the Nss is well above the recommended 11,000 limit, and the NSPAH margin ratio is too low for the calculated suction energy.

TABLE 3.3
NPSH MARGIN RATIO GUIDELINES (NPSHA/NPSHR)

SUCTION ENERGY LEVEL	
Suction Energy Level	Margin
Low	1.1-1.3
High	1.3-2.0
Very high	2.0-2.5

Threshold Values for End Suction Pumps are defined as:
High suction energy 160 x 10^6
Very high suction energy 240 x 10^6

Now let's try the analysis at 1780-rpm in hopes of lowering the N_{ss} value. The N_s value of 503 at 3580 rpm suggests that can achieve a decent hydraulic efficiency of about 64 percent at BEP). If you consider this efficiency too low, you can increase the number of stages until the N_s reaches the range between 1500 and 2500, where hydraulic efficiency peaks.

The estimated efficiency values can be extracted from the graph similar to the one found in Figure 3.3. You simply look up expected hydraulic efficiency based on N_s and the rated flowrate. The estimated efficiency can be determined by first finding the specific speed calculated by the spreadsheet along the horizontal axis, traveling up until the rated flow curve is found and then moving horizontally until you reach the vertical axis. For example, a pump with a specific speed of 1000, flowing 1000-gpm will have an efficiency of about 75 percent. Once you input the value of expected efficiency the required brake horsepower is calculated.

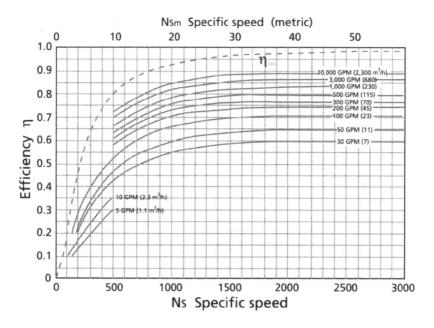

Figure 3.3 - Efficiency vs. Specific Speed (based on BEP flow)

Table 3.4 shows the results of the same process requirements examined at 1780-rpm.

Table 3.4- Pump Design Analysis Table (1780 rpm Pump)

Inputs	Units	
RPM		1780
Flow @ BEP	gpm	500
Suction pressure	psi	25
Discharge pressure	psi	380
Specific gravity		0.95
Pump efficiency		0.79
NPSHa	ft	12
NPSH margin	ft	2
Number of stages		10
Number of suction eyes		1

Outputs	Units	
NPSHr	ft	10
NPSH ratio		1.2
Pump head	ft.	863.21
Head per stage		86.32
Ns		1405.46
Nss		7077.91
Impeller suction eye diameter (estimate)	inches	6.00
Suction energy		7.18E+07
Brake horsepower	HP	131.07

Notice that by dropping the speed down to 1780 rpm the N_{ss} drops to an acceptable 7078 for a 10-stage design and that the N_s rises to 1405 (somewhere between a Radial vane and a Francis vane impeller design), which raises the estimated efficiency to 79 percent (at BEP). This efficiency improvement represents a horsepower savings of about 30.7-hp, or a 23.4 percent reduction in energy costs over the pump's lifetime.

(Note: At an electrical rate of $0.08/Kwh and inflation rate of 3 percent, this equates to an annual savings of $16,000 or $234,000 over a 20-year pump life.) The analysis also reveals that the suction energy dropped from 290 x 10^6 down to 71 x 10^6, which places it in the low suction energy range and allows a NPSH margin ratio in 1.1 to 1.3 range (see Table 3.5). Since we only have a NPSH margin ratio of 1.2, you might want to see if you can increase the NPSHa slightly so that it exceeds 1.3.

Here's one final example.

Table 3.5- Pump Design Analysis Table (3580 rpm Pump); 1000 gpm and a 225 psi differeintiual.

Inputs	Units	
RPM		3580
Flow @ BEP	gpm	1000
Suction pressure	psi	25
Discharge pressure	psi	250
Specific gravity		0.95
Pump efficiency		0.78
NPSHa	ft	40
NPSH margin	ft	15
Number of stages		1
Number of suction eyes		1

Outputs	Units	
NPSHr	ft	25
NPSH ratio		1.6
Pump head	ft.	547.11
Head per stage		547.11
Ns		1000.76
Nss		10125.77
Impeller suction eye diameter (estimate)	inches	6.00
Suction energy		2.07E+08
Brake horsepower	HP	168.27

After inputting all the design requirements, we get an Ns of 1000 and an Nss if 10,126. From Figure 3.3, you see that you can expect an efficiency of 78%. Overall this looks like a centrifugal pump is an excellent choice for this pump application. The Ns value is in the middle of the desired Ns range and the Nss value is comfortably below 11,000. You also see that the suction energy value of the 207×10^6 points to a high suction energy level, which requires an NPSH margin ratio in the 1.3 to 2.0 range. You should be fine with a margin of 1.6, but If possible we should consider finding a way to increase the NPSHa to achieve an NPSH margin ratio of 2.0.

These relatively simple calculations enable you to predict what an ideal pump would look like and quickly determine what the critical input parameters are. If the N_{ss} value is too high, consider slowing the pump down to see if you can drop it into acceptable range. If your N_s value is too low, simply add stages to raise the value into a range where the efficiency is tolerable.

Tip: What if you are trying to determine the NPSH margin requirements using the suction energy method, but don't have any impeller geometry details? If you are in the initial phases of your pump selection process, you likely don't have any impeller specifics. What can you do?

You can estimate the impeller eye diameter with the following formula in order to calculate the suction energy and then determine the recommended NPSH margin ratio:

$$D_{eye}\,(inches) = \sqrt{0.037 \times Q(gpm)}$$

In this formula, Q is the pump design flow (in gpm) and D_{eye} is the impeller eye diameter (in inches). The impeller eye diameter estimation formula assumes an average fluid velocity of 11 ft/sec at the impeller eye. When you have an actual impeller eye diameter from the manufacturer, you can insert it into the suction energy formula to ensure you have a sufficient NPSH margin.

Armed with these simple analysis tools, you can talk to the process engineer to determine if the tower or liquid control level can be raised to improve the NPSH$_a$, or if the pump control valve pressure drop can be lowered slightly to eliminate the need for an extra stage. You can also use this worksheet to sell the merits of lower speed operation to the project engineer. This worksheet won't solve all your problems, but it will allow you to think more like a pump designer before going out for quotations. Pre-selecting the type of pump you would like will make the pump selection process more efficient and less frustrating.

Some advice

Our biggest pump selection challenge is fighting the centrifugal pump paradigm. When you mention a centrifugal pump in a process setting, most people usually envision a single stage, end suction pump, operating at 3600-rpm. Why is this? Because this design requires only one mechanical seal and usually has the lower purchase price when compared to an 1800-rpm design. However, 3600-rpm pumps can come with a price in the form of less process flexibility or mechanical unreliability. They usually require either higher NPSH$_a$ values or higher N$_{ss}$ impeller designs. The adverse effects of higher N$_{ss}$ designs were well documented in the 1980s. This led to stricter guidelines for Nss values. Now that most pump manufacturers limit their Nss designs to less than 11,000, flow instability problems are becoming less common. Now 3600-rpm pumps are considered the norm.

We all this being said, you shouldn't automatically consider 3600 rpm, single stage pumps as the go-to design configuration for all our applications, even though the people controlling the money like their economic benefits. I challenge you to first conduct a thorough analysis of the hydraulic requirements as a means of exploring all viable hydraulic choices before narrowing down your selection field.

Chapter Summary

In this chapter you have learned how the hydraulic parameters Ns and Nss can be useful in the pump selection process. Remember to keep these rules of thumb in mind:

1. Hydraulic efficiency peaks at specific speeds (N_s) between 2000 and 3000 (in US units) and drops dramatically below 500 (in US units). If efficiency is important, try to select pumps with specific speeds in the 2000 to 3000 (in US units) range.
2. Only select pumps with suction specific speeds (N_{ss}) less than 11,000 (in US units)—an N_{ss} of less than 9000 (in US units) is even better.

To succesfully select the perfect pump, the pump engineer must look at the whole picture, which includes the suction piping system, the process design requirements, the overall goals of the project engineer and the maintenance engineer.

Caution: Do not work in a vacuum. Today, employers highly regard employees who work well in team settings. We all need to work together and pull together—hopefully all in the same direction. I am hoping this pump analysis calculator will enlighten and encourage readers to better understand their pumps and select their pumps for the right reason. By thinking like a pump designer you will be better equipped to:

- Develop better pump specifications.
- Talk to the suppliers of your pumps with the goal of improved pump selection.

Chapter 4
10 Pump Selection Rules of Thumb

The key to exemplary centrifugal pump reliability and high overall system efficiency is selecting the right pump for the required application. This means that the pump's best efficiency flow (BEP) and head should always closely match process' hydraulic requirements. A poor hydraulic match will lead to high energy bills and maintenance costs.

Centrifugal pumps like to operate close to their best efficiency point (BEP) where:

- Hydraulic efficiency is greatest
- Vibration and pressure pulsations are lowest
- Shaft deflection is minimized due to the improved pressure distribution of internal pressure around the impeller. This has a major positive effect on mechanical seal life. At the best efficiency point, the pressures around the impeller are nearly balanced so that radial forces on the impeller and shaft are minimized.
- Overall mechanical reliability is highest

BEP is the flow point at which centrifugal pump designers would like to see their pumps operating. Conscientious pump operators should always strive to comply with the designer's intent by operating all centrifugal pumps near their best efficiency point.

10 tips to selecting efficient and reliable centrifugal pumps

If you are not a pump specialist, selecting the right centrifugal pump may seem to be a daunting task. There are countless designs and features to choose from. There are: 1) horizontal and vertical pumps, 2) single stage and multistage pumps, 3) overhung and between bearing pumps, etc. Where to start?

To make the pump selection process easier, I have summarized some key aspects of centrifugal pumps along with some recommendations based on years of experience. By following the recommendations listed below, you should stay clear of major problems and select efficient and reliable pumps for your given applications.

1. **Only select pumps with suction specific speeds (N_{ss}) less than 11,000 (in US units)—an N_{ss} of less than 9000 (in US units) is even better.**

   There have been numerous technical papers and articles published reporting suction recirculation problems in pumps with high suction specific impellers. One of these papers, written by Hallam (1) in 1982, included actual failures data from a 5 year study of 480 centrifugal pumps in Amoco's Texas City refinery. Most of these pumps were in hydrocarbon services. These pumps ranged in size from 150 hp to a maximum of 1,000 hp. Hallam's data clearly illustrated that the failure rate of pumps with suction specific speeds above 11,000 was almost twice that of those with suction specific speeds less than 11,000. The apparent explanation for these findings is that pumps with high suction specific speeds tend to be hydraulically unstable when operated away from their design point. Eventually, 11,000 became the accepted limit for suction specific speed for pumps in process applications.

To ensure there is adequate net positive suction head available (NPSH$_a$) for proper pump operation, we need to know the net positive suction head required (NPSH$_r$) for the pump being considered. Insufficient NPSH available may seriously restrict pump selection, or even force an expensive system redesign. On the other hand, providing excessive NPSH available may needlessly increase system cost.

The suction specific speed parameter may provide help in this situation by understanding the trade-offs between different pump parameters. Suction specific speed (S) is defined as:

$$N_{ss} = \frac{N \times (GPM)^{1/2}}{(NPSH_R)^{3/4}}$$

Where
N = Pump speed RPM
GPM = Pump flow at best efficiency point at impeller inlet (for double suction impellers divide total pump flow by two).
NPS$_R$ = Pump NPSH required at best efficiency point.

Example - Specific Suction Speed

The available Net Suction Head – NPSH$_R$ of a pump is given to be *20 ft*. With a rotational speed of *1750 rpm* and a flow rate of *500 US gpm,* the specific suction speed is calculated to be:

N_{ss} = *(3550 rpm) (500 gpm)$^{1/2}$ / (20 ft)$^{3/4}$ = <u>8393.5</u>*

This is below the high limit of 11,000 and even below the low limit of 9000.

2. **Never select a pump that will have to operate continuously below 70% to 80% of its best efficiency point.** Remember that during start-ups, shutdowns, and upsets pumps tend to deviate from their design conditions. So, if the pump starts at 70% of the BEP, it may end up operating at 40% or 50% of the BEP during off-design conditions. This will dramatically affect the pump's useful life.
3. **Remember that 1800 rpm (or 1500 rpm) and slower pumps are usually more reliable that 3600 rpm (of 3000 rpm) pumps.** Don't rule out slow speed options simply because they are more expensive. Factor in the long term benefits of higher efficiency and lower maintenance into your final decision.
4. **Hydraulic efficiency peaks at specific speeds (N_s) between 2000 and 3000 (in US units) and drops dramatically below 500 (in US units). If efficiency is important, try to select pumps with specific speeds in the 2000 to 3000 (in US units) range.**

Specific speed (N_s) is a non-dimensional design index used to classify pump impellers based on their type and proportions. N_s is defined as the speed in revolutions per minute at which a geometrically similar impeller would operate if it were of such a size as to deliver one gallon per minute against one foot head. The following formula is used to determine specific speed:

$$N_s = \frac{N \times \sqrt{Q}}{H^{3/4}}$$

Where
N = Pump speed in RPM
Q = Capacity in gpm at the best efficiency point
H = Total head per stage at the best efficiency point

Figure 4.1 below depicts the general relationship between specific speed and efficiency. As a rule of thumb, we can say that pump efficiency typically peaks at N_s values between 2000 and 3000.

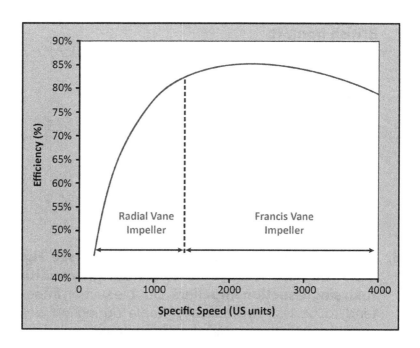

Figure 4.1, General relationship between
pump efficiency and specific speed
(This graph is for illustration purposes only. Actual pump
efficiencies should be provided by manufacturers.)

Example - Specific Speed for a Pump - with different units

A pump has a capacity of *1500 US gal/min (1249 British gpm, 340 m³/h, 94.4 liters/min)* at *100 ft (30.5 m)* of head and is rotating at *1760 rev/min.* Specific speed can be expressed as follows:

US gpm, ft:
$N_{s(US\ gpm,\ ft)}$ = *(1760 rev/min) (1500 gal/min)$^{1/2}$ / (100 ft)$^{3/4}$ = 2156*

British gpm, ft:
$N_{s(British\ gpm,\ ft)}$ = *(1760 rev/min) (1249 gal/min)$^{1/2}$ / (100 ft)$^{3/4}$ = 1967*

m³/h, m:
$N_{s(m3/h,\ m)}$ = *(1760 rev/min) (340 m³/h)$^{1/2}$ / (30.5 m)$^{3/4}$ = 2500*

liter/min, m:
$N_{s(liters/min,\ m)}$ = *(1760 rev/min) (94.4 m³/h)$^{1/2}$ / (30.5 m)$^{3/4}$ = 1317*

5. **Use double suction impellers sparingly (see Figure 4.2). They are less stable at off-design conditions than single suction impellers.** Double suction impellers tend to be less hydraulically stable during off design conditions. Only consider double suction impellers if the flow is expected to be at or near BEP conditions.

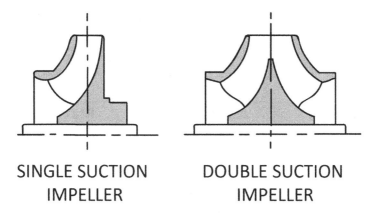

SINGLE SUCTION   DOUBLE SUCTION
IMPELLER   IMPELLER

Fig. 4.2- Comparison of a single suction impeller
(left) and a double suction impeller (right)

6. **Never select pumps with a maximum diameter impeller. You may need to increase the impeller diameter in the future for more flow or head.**

7. **Always provide expected normal, minimum, and maximum pumping rates and temperatures in the bid specifications. This will allow bidders to make pump and seal recommendations that will meet the true process need**s. (Note: If known, also provide startup and upset pump flow and temperature information as additional information on the pump data sheet.)

8. **Use hydraulic stability, not temperature, rise as criteria for setting the minimum acceptable pump flow.**

   As mentioned above, pumps with higher suction specific speed numbers tend to have narrower stable operating flow ranges. This must be taken into account when setting the stable flow limits of a centrifugal pump. Table 4.1 below provides basic guidelines for minimum and maximum flow based on the pump's suction specific speed. For example, a pump with a suction specific speed of 10,000 can expected to be operated reliably from 77% to 109% of its best efficiency point.

Table 4.1, Guidelines for Stable Pump Operation

Suction Specific Speed (US units)	Minimum recommended flow for stable operation (in % BEP flow)	Maximum recommended flow for stable operation (in %BEP flow)
5000	65%	118%
6000	67%	117%
7000	70%	116%
8000	72%	114%
9000	75%	113%
10000	77%	111%
11000	81%	109%
12000	85%	108%
>12000	Consult OEM	Consult OEM

If your calculations indicate your pump will operate below the recommended minimum flow for a significant length of time, you will need to install some type of spillback line for pump protection. If your calculations indicate your pump will operate above the recommended maximum flow, you will need to start a second pump in that situation to avoid damage.

9. **Incorporate a healthy NPSH (net positive suction head) margin or ratio, i.e. $NPSH_a/NPSH_r$, into your selection. This ratio should be anywhere from 1.1 to 2.0 depending on the liquid, criticality, and suction energy level. A larger NSPH margin is always better.**

10. Consider liquid volatility when making your pump selection. Be more conservative in your pump selection when the liquid has a single boiling point (more volatile), as opposed to a liquid with a wide boiling point range (less volatile).

Here are some special design recommendations to consider when making your final selection:

- Pumps that are to be piped in series must have the same capacity (impeller width and speed)
- Pumps that are to be piped in parallel must have the same head (impeller diameter and speed)
- Use double volute pumps any time your impeller diameter is 14 inches (355 mm) or greater.
- Specify centerline support pump designs when the pumping temperature exceeds 200 degrees Fahrenheit (100° C). This design will allow the wet end of the pump to expand in two directions, better maintaining pump to motor alignment, as well as the internal pump-end internal clearances.
- Never select a pump that is specified with an impeller that exceeds 95% of the maximum diameter. You might need to install a larger impeller diameter for some future process requirement.
- Consider a variable speed pump driver if you have a friction-controlled system. Pipeline or circulation pumps are examples of friction-controlled pumping systems. If you have a high static differential head, as is the case with a boiler feed pump, a variable speed design will not be of much use in keeping the pump near the best efficiency point.
- High energy pumps, defined as pumping to a head greater than 650 feet (198 meters) and more than 300 horsepower (224 KW) per stage, require special consideration to avoid blade passing frequency vibrations and low frequency vibrations at reduced flow rates.

Team Approach to Pump Selection

A team approach to pump selection process is highly recommended. No one person has all the information, knowledge, or experience required to make the appropriate pump selection decision every time. To improve your odds of making the right decision, you need to engage key site specialists with a vested interest in the long term performance and reliability of the pump. As a minimum, I recommend that a team composed of a project engineer, process engineer, operator, machinery engineer, and maintenance professional be involved in pump bid review and selection process. Remember that the "optimum" pump selection is not the cheapest pump, but the one that represents the lowest cost of ownership over its service life (2). When reviewing bid tabs for critical pumps, it might make sense to bring in the pump vendors individually to answer any final design questions in person. This effort will pay off with a reliable and efficient pump operation for years to come.

Chapter 5
Bid Tabulations

There comes a time in all projects when the rubber has to hit the road. We can perform numerous calculations based on design conditions to better understand the hydraulic requirements of the pump in question. But eventually, we need to determine if there is a commercially available pump that can actually fit the design requirements.

Remember that you have started your pump selection quest by using specific speed calculations to determine if a centrifugal pump will work for your application and the suction specific speed calculations to see if there is adequate NPSHa available. First, by assuming you would be operating at the best efficiency flow, you determined if the specific speed fell in the 1000 to 6000 range. If you passed the Ns test, you proceeded to the theoretical Nss calculations to see if the Nss fell below 11,000 assuming you are operating at the best efficiency flow.

If your application passes the preliminary Ns and Nss tests, then the next step is to submit your hydraulics specification to various pump manufacturers to see if they offer pump designs that can fit your requirements. During the preliminary calculations, we made a <u>big</u> assumption that our selected pump will be operating at its best efficiency point flow. In reality, this is rarely the case.

After the manufacturers (or vendors) have had a chance to review your hydraulic requirements, they will either 1) not bid if they don't have a pump that meets your requirements, 2) bid a single pump, or 3) bid two or more options. It can be overwhelming if your purchasing department sends you emails with numerous attachments full of bidding information. How do you condense all the information so that it is manageable? You can simplify your evaluation process by compiling the manufacturer's data into a document called a bid tabulation (also called a bid tab).

Table 5.1 is an example of pump a bid tabulation table. Notice that the bid tabulation data is arranged in columns. There is a column for each pump that is offered as an option. Next, notice that there are rows for key pump evaluation parameters. Here is a short discussion of the evaluation parameters:

1. %BEP at rated flow is a measure of how good a fit the pump is for your application. You want your %BEP of rated flow to be, at the very least, above 60% and hopefully over 80%. In the pump curve in Figure 5.1, the best efficiency flow is at 1000 gpm, which means a design flow of less than 600 gpm would represent a %BEP flow of less than 60%. Similarly, a design flow of less than 800 gpm would represent a %BEP flow of less than 80%. Remember that the BEP flow is the flow on the pump performance map where the efficiency peaks.
2. Nss is the actual suction specific speed for your application. You should never accept an Nss greater than 11,000.
3. Horsepower that is listed at the maximum flow lets you know what size driver will be required. As the pump efficiency drops, the horsepower increases and annual power costs go up.

4. % of full diameter impeller indicates if the impeller diameter can be increased in the future in the event additional flow is required. Obviously if you are at 100% of the full diameter, there is no room for uprating. However, if you are at 85% of the full diameters, there is plenty of room to increase the impeller diameter for more head or flow.

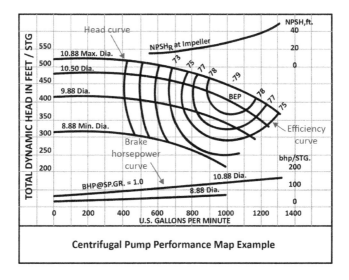

Fig. 5.1 -Typical Pump Curve. The BEP flow is 1000 gpm.

Knockouts

Knockouts are bid tab parameters that cause the pump to be immediately eliminated from the competition. Parameters such as %BEP flow, Nss, % of full diameter impeller are examples of bid tab values that can knock a pump out of the running.

- Any BEP flows that are less than 60% should drop the pump from the competition
- An Nss greater than 11,000 should drop the pump from the competition
- Any pump with an impeller greater than 95% of its full diameter should not be considered

Efficiency could potentially be a knockout depending on the scope of the project you are working on. In large (>1000hp) pumps that have a 10% or 15% lower efficiency ranking represents a significant increase in energy costs. In contrast, energy costs due to efficiency differences are less significant in pumps less than 5 hp.

Bid Tabulation Example #1

Let's review the following bid tabulation. Below we have a bid tab for Main Column Bottoms Pumps (P100A & B). The upper table summarizes the basic design requirements. The lower table lists all the key pump selection parameters for four different pump options. The cells that are shaded gray are considered knockouts. For example, all the pumps are knocked out except Pump A because of the prediction that they will experience flows below 60% at their minimum flows. Also Pump B is eliminated because it has an Nss of 12,037 and a % of maximum impeller diameter of 100%. This is a relatively simple bid tab, because there is only one pump left standing, i.e. Pump A.

Table 5.1 Bid Tab Summary #1

Specific gravity	0.85	
Minimum flow	500	gpm
Maximum flow	800	gpm
Rated flow	700	gpm
Rated head	500	ft
NPSHa	45	ft

	Pump A	Pump B	Pump C	Pump D
BEP flow	750	1000	1200	1400
%BEP at min flow	66.7%	**50.0%**	**41.7%**	**35.7%**
%BEP at max flow	106.7%	80.0%	66.7%	**57.1%**
%BEP at rated flow	93.3%	70.0%	58.3%	**50.0%**
NPSHr (at BEP)	25	20	30	15
Nss	8818.16	**12037.33**	9728.64	8836.25
NPSHr at max flow	26.67	16.00	20.00	8.57
NPSH ratio	1.6875	2.8125	2.25	5.25
NPSH margin (ft)	18.33	29.00	25.00	36.43
% of full diameter impeller	80.0%	**100.0%**	85.0%	95.0%
Eff at rated flow	75.0%	70.0%	**55.0%**	**50.0%**
HP at min flow	71.55	95.40	114.48	133.56
HP at max flow	114.48	122.66	156.11	171.72
Rated speed (rpm)	3600	3600	3600	1800
Number of stages	1	1	1	2
Cost	$ 22,000	$ 28,000	$ 25,000	$ 35,000

Bid Tabulation Example #2

Here is another bid tabulation example for Water Circulation Pumps (P2001 A & B) (Table 5.2). As before, the upper table summarizes the basic design requirements. The lower table lists all the key pump selection parameters for four different pump options. The cells that are shaded gray are considered knockouts. For example, Pumps C and D are knocked out because the manufacturers predict that they will experience flows below 60% at their minimum flows. Also, Pumps A and D are eliminated because they have Nss values above 11,000. As in example #1 there is one pump left standing. In this case Pump B is the only viable option.

Table 5.2 Bid Tab Summary #2

Specific gravity	1	
Minimum flow	1000	gpm
Maximum flow	1500	gpm
Rated flow	1250	gpm
Rated head	350	ft
NPSHa	52	ft

	Pump A	**Pump B**	**Pump C**	**Pump D**
BEP flow	1500	1400	2500	3000
%BEP at min flow	66.7%	71.4%	**40.0%**	**33.3%**
%BEP at max flow	100.0%	107.1%	60.0%	**50.0%**
%BEP at rated flow	83.3%	89.3%	50.0%	**41.7%**
NPSHr (at BEP)	25	42	30	15
Nss	**12470.77**	8164.50	7021.04	**12934.95**
NPSHr at max flow	25.00	45.00	18.00	7.50
NPSH ratio	2.08	1.16	2.89	6.93
NPSH margin (ft)	27.00	7.00	34.00	44.50
% of full diameter impeller	80.0%	90.0%	**100.0%**	95.0%
Eff at rated flow	75.0%	72.0%	**55.0%**	**50.0%**
HP at min flow	117.85	109.99	196.41	235.69

HP at max flow	176.77	184.13	241.05	265.15
Rated speed (rpm)	3600	3600	1800	1800
Number of stages	1	1	2	2
Cost	$ 44,000	$ 42,000	$ 55,000	$ 58,000

In cases where there is more than one viable option that survives the bid tabulation analysis, you must find a way to force rank the remaining pump. Parameters as efficiency, cost, % of full diameter impeller, or experience with the manufacturer are a few factors than can be used to force rank the remaining options. Once you have completed the bid tabulation process and forced rank the viable options, it will be easy to sell your final selection to the project manager and the other stakeholders.

In the event that all the bids are rejected, you will have to revisit your hydraulic assumptions and determine if any of them can be modified. For example, you may need to increase a tower or sump level to increase your NPSHa or you might have to reconsider your true flow range requirement. You might also have to work with your manufacturers to find the right combination of design requirements to allow a reliable pump selection options.

Chapter 6
The most important day in a pump's lifetime

A few years ago I saw a commercial on TV touting an air condition system, where the announcer stated, "The most important day of an air conditioning system's life is the day it's installed." I thought about this a lot and decided that this is not true for pumps, where there are many important days in a pump's life. If you want a perfect pump installation here are days I think are truly critical:

- The day pump hydraulic requirements are specified by the process engineer
- The day the pump specifications are written
- The day the pump is selected
- The day the mechanical seal is selected
- The day the piping and control system are designed
- The day the pump is installed

Additionally, if you want to reap the benefits of your "near perfect" pump installation, you must also be aware of these important days:

- ◦ Every day it's repaired
- ◦ Every day it's started up

In reality, there can never be a perfect pump installation. Process engineers never know exactly what flow range or pressure rise will be required during the life of the pump. Processes tend to evolve due to market forces and technological improvements, forcing pumps to operate away from their sweet spots. Also, pump manufactures don't always have the exact pump to fit your needs. To make things worse, tight project economics can severely limit the level of design features your installation is allowed to have.

Thinking about all these selection forces and considerations can give a pump user a severe case of anxiety. How can we ever hope to obtain a perfect pump installation? There is no definitive answer to this impossible question. All I can provide are a few guidelines based on many years of experience in the pump trenches:

1. Have the process engineer specify the possible range of flow and head requirements, not just maximums. Minimum flow rates for start-ups or other special conditions may require spillback lines or variable speed drives for pump protection.
2. Have the process engineer specify the possible range of process temperatures and if the stream is expected to be dirty or corrosive. This input will allow for a suitable mechanical seal selection and flushing plan.
3. Follow PIP guideline RECP001 (Design of Pumping Systems That Use Centrifugal Pumps) when designing the pump's piping system. Also diligently follow PIP guideline REIE686 (Recommended Practice for Machinery Installation and Installation Design) to ensure

your pump is installed relatively stress free and firmly grouted in.

4. Select a control system that allows your pump to operate at the lowest stress level, i.e. near the best efficiency point, most of the time. Beware of temperature and level controls that can force your pump to operate well away from the pump's BEP.

After your pump has been purchased and installed, remember that the road to pump reliability must run through your training department. You must train your operators to understand what pump reliability is all about and how to put their knowledge into practice. They need to understand that every start-up, shut-down, and inspection is an opportunity to improve their process' reliability.

This goes for your mechanics also. Instill in them that they are not just parts changers, they are also reliability technicians. Every pump failure and subsequent repair is an opportunity for improvement. Frequent discussions between the operators, mechanics, and machinery engineers should be encouraged as a means of solving problems related to detrimental operating and repair practices.

I'll leave you with this quote attributed to Admiral Hyman G. Rickoever, U.S. Navy, known as the "Father of the Nuclear Navy".

"The devil's in the details—but so is salvation"

Details matter, especially during the most crucial days of a pump's life. Remember that pump reliability is everybody's job, not just the machinery engineer's.

Chapter 7
The Pump Selection Process

Now we will tie all the concepts in this book together and summarize the centrifugal pump selection process. Here are seven steps that will allow you to go from a set of design specifications to the final selection of a reliable pump.

Step1: Get all the pump hydraulic design requirements from the process engineer. Preferably, have the process engineer enter the data into a sheet similar to the one shown in Figure 7.1. Entering the data into a spec sheet ensures that you have all the required selection data.

Best Flow Pumps Sales and Service	Horizontal and In-Line Centrifugal Pump Data Sheet	Specification Number:	
	Client: XYZ Chemicals	Sheet 1 of 1	

Number: 1	Service: PIPE LINE BOOSTER	Item No.:	1
Manufacturer: Ace Pumps	Size / Type: 6x8x11 DMSD-10(9)	E/R No:	2
Motors: Required 1 Provided By: BFP Mounted By: BFP		Item No: J-3132	3
Tag Number: P6100			4
OPERATING CONDITIONS, EACH PUMP		**PERFORMANCE**	5
Liquid: LIGHT HYDRO CARBON US GPM at Point: Norm: 1110 Rated: 1162		Proposal Curve No.: MD-84/MD-131	6
Discharge Pressure, PSIG: 1255		Stages: 4 RPM: 3560 at 60 Hz	7
Pt. F, Norm: 75 Max:	Suction Pressure, Max, PSIG:	NPSHR (Water): 28 ft	8
S.G. at Pt.: .489	Suction Pressure, Rated, PSIG: 430	Efficiency: 79% Rated BHP: 747	9
Vap. Press at PT, PSIA 350	Diff Pressure: 805 PSI 3802 ft.	Max HP Rated Impeller: 10 7/8"	10
Viscosity at PT 75 cP NPSHA 542 ft. Hyd. HP 545.7		Max Head Rated Impeller: 4635 ft	11
Corrosion / Erosion by:		Stuffing Box Pressure: 540 PSI	12
		Min Cont. GPM: 450	13
CONSTRUCTION		Rotation Facing Coupling CCW	14
			15

	Size	Rating	Facing	Location	SHOP TESTS	
						16
Suction	8	900	RTJ	SIDE	O Perf. Test O Witnessed O Non-Wit	17
Discharge	6	900	RTJ	SIDE	● Hydro-Test O Witnessed ● Non-Wit	18
Thrust: lbs O Up O Down					O NPSH Test O Witnessed O Non-Wit	19
Case Mount: O Centerline ● Near Centerline O Foot O Bracket O Inline					O Vibration O Witnessed O Non-Wit	20
Case Split: ● Axial O Radial Volute: O Single ● Double O Diffuser					O Shop Inspection	21
Max Allowable Pressure: Suction 2000 PSIG; Discharge: 2800 PSIG; 100 F					O Dismantle and Inspect After Test	22
Hydro: 2400 PSIG; Held 35 minutes						23
Aux. Connections / Size: ● Vent 1/2 ● Drain 1/2" ● Gauge 1/2" O					**MATERIALS AND ACCESSORIES**	24
Impeller Diameter: Rated: 10 7/8" Max: 11 1/2" Type: ENCLOSED					Case / Trim: API Column S-5	25
Mounted: ● Between Bearings O Overhung					Case: CARBON STEEL	26
Bearings, Type: Radial SLEEVE Thrust: SLEEVE / KTB					Impeller: CARBON STEEL	27
Lube: O Ring Oil O Flooded O Oil Mist O Flinger ● Pressure					Shaft: AISI 4140 c/s	28
● Coupling: Mfr.: METASTREAM Model: TSA6-0400-7"					Shaft Sleeve: 18-8 SS	29
Driver half mounted by: Best Flow Pumps					Case Wear Ring: 12% CHROME	30
● Mechanical Seal: Mfr.: BORG WARNER Type: QBQ-3750/3500 API Code:					Throttle Bushing: N/A	31
Description: TANDEM MATERIAL SU4X					Gland: 316 SS	32
● Baseplate: OIL FIELD - SEE GA-3132-01 & GA-3132-02					Gaskets: NON-ASBESTOS	33
● BEARING RTDs: 3) 100Ω PLATINUM - MINCO						34
● VIBRATION SWITCH: PMC/BETA - SEISMIC TYPE						35
AUXILIARY PIPING					Base Plate: STEEL	36
O Cooling Water Plan: Material:	O Tubing O Piping				Coupling Guard: STEEL	37
● Primary Seal Flush Plan: 11 Material: 316 SS -.065 ● Tubing O Piping					Coupling: STEEL	38
● Secondary Seal Flush Plan: 52 Material: 316 SS -.065 ● Tubing O Piping					Spacer: STEEL	39
O						40
C.W. Requirements: O Water Cooling O Pedestal						41
O Bearings O Stuffing Box					Weights: Pump: 8810 lbs	42
O Gland O					Motor: 7500 lbs	43
O TOTAL NONE Pressure / Temp					Base: 7700 lbs	44
					Total: 24010 lbs	45
MOTOR SECTION						46
Item No.: Mounted By: CUSTOMER					**MFR FINAL DATA (AS BUILT)**	47
HP: 800 RPM: 3600 Frame: 30RS6					Actual Impeller Dia: 10 7/8"	48
Manufacturer: ALLIS CHALMERS					Test Curve No.:	49
Type: INDUCTION Insulation: B Enclosure: TEFC					Seal Dwg. No.:	50
Volts/Phase/Cycles: 460-3-60 (INVERTER DRIVE)					Pump Serial No.: 34220393	51
Bearings: SLEEVE Lube: OIL Temp. Rise: 60					Pump Drawing No.: GA-3132-01	52
Full Load Amps: 895						53
						54
API 610 Applies O Yes ● No Edition:						
Revision Number:						
Prepared By/Date						
Checked By/Date						
Auth. By/Date						

Fig. 7.1 – Typical Pump Specification Sheet

Before proceeding with the preliminary hydraulic pump calculations, make sure that:

1. A thorough system hydraulic analysis was conducted so that the design head required for the pump is accurate and accounts for all the system elements. A system head curve is preferred, but a single point is the minimum requirement for analysis.
2. There is not too much entrained gas or vapor in the fluid to be pumped. Remember that centrifugal pumps are

only efficient if you are handling liquids containing less than approximately 2% vapor.

3. Solids in liquid are small enough to pass through the pump impeller passages. As a rule of thumb the solid particles should be less than half of the smallest impeller opening.
4. The fluid viscosity is less than 1000 cSt. If the viscosity is greater than 1000 cSt, make sure to consider the economics related to the higher pumping energy cost.

If you haven't conducted a hydraulic analysis of your pumping system, make sure it's complete before proceeding. If any of last three factors are not true, you should reconsider using a centrifugal pump.

Step 2: Calculate the ideal Ns and Nss values using the design data from the data sheet. Use the design flow, head and speed values in your calculations. (Notes: 1) The "design" values of flow and head are the points the customer wants the manufacturer to guarantee. The pump generally will not operate continuously at its design or rated point, but it is designed and tested by the manufacturer with this performance guarantee in mind. 2) When using the Nss formula to calculate a preliminary Nss value, set NPSHr equal to NPSHa plus a healthy NPSH margin, i.e. NPSHr = NPSHa + NPSH margin.)

Step 3: Check to see that the Ns and Nss values are acceptable, i.e. Ns>500 and Nss<11,000.

Step 4a (You find the Ns and Nss values to be acceptable): Go back and ensure that the NPSH margin ratio if acceptable now that you know the Nss value is <11,000. If the NPSH margin ratio is acceptable, you can proceed to step 5.

Step 4b (You find the Ns and Nss values to be unacceptable):

- If the Ns is too low, consider increasing the number of stages or increasing the rotation speed. As a rule of thumb, you should consider 3600 rpm as the upper limit for most applications. Note: If you can't find a combination of impeller stages or speed that will increase the Ns value above 500, consider using a plunger, screw, or diaphragm pump.

• If the Nss value is above 11,000, consider either increasing the NPSHa or using a vertical turbine pump. A vertical turbine pump places the first stage impeller well below the suction flange to gain suction head. Another option is to decrease the rotational speed. Talk over the various options with the process engineer to see which one makes the most sense. If you discover there is no way to satisfy the Ns and Nss requirements, you may have to consider another type of pump.

Step 5: If you find that your pump requirements are well suited for a centrifugal pump selection, you can proceed with requesting pump bids from qualified manufacturers. It is best to have at least 3 to 4 manufacturers bid if possible, however the absolute minimum is two bidders to obtain competitive quotes. It is acceptable for manufacturers to offer more than one pump option. They will sometimes bid an alternative option to increase their chances of winning the bid.

Step 6: Perform detailed bid tabulation to enable you to objectively evaluate all the bids. It is best to list the top three selections in order of best to worst after you have completed the bid analysis. If for some reason your first choice drops out or is disqualified, you will have your other options identified.

Step 7: The final step is to sit down with the process engineer, maintenance engineer, the purchasing agent, and any others with vested interests, to make the final pump selection.

After you have conducted a thorough hydraulic design and bid tab analysis and allowed everyone to add their inputs, you can be sure to make the best pump selection based on the available information. Now all you have to do is to order the pump, if only one is required, or pumps, if there is a main and spare pump required for the application.

Appendix A

How to Calculate Pipe Friction Head Losses for Newtonian Fluids

By
Jacques Chaurette
Mechanical Engineer

To construct an accurate system head curve, we must be able to determine the friction head losses throughout the entire piping system. The friction head is the fluid energy losses (in feet or meters) due to the movement of fluid in a piping system and is proportional to flow rate, pipe diameter and viscosity. The friction head, as defined here, is made up of the cumulative fluid energy losses due to the fluid movement and the friction loss due to the effect of pipe fittings (for example, 90° elbows, 45° bends, tees, etc.):

$$DH_F = DH_{FP} + DH_{FF}$$

The subscript FP used here refers to pipe friction loss and the subscript FF refers to fittings friction loss.

Newtonian Fluids

Newtonian fluids are a large class of fluid, whose essential property, viscosity, was first defined by Newton. Viscosity is the relationship between the velocity of a given layer of fluid and the force required to maintain that velocity. Newton theorized that for most pure fluids, there is a direct relationship between force required to move a layer and its velocity. Therefore, to move a layer at twice the velocity requires twice the force. His hypothesis could not be tested at the time, but was later demonstrated to be valid by the French researcher, Poiseuille. His research resulted in a very practical definition for viscosity.

The Darcy-Weisbach formula expresses the resistance to movement of any fluid in a pipe:

$$\frac{\Delta H_{FP}}{L} = f \frac{v^2}{D \times 2g}$$

where f is a non-dimensional friction factor. Often, the tables give values for friction loss in terms of feet of fluid per 100 feet of pipe. When the appropriate units are used (Imperial system), the Darcy-Weisbach equation becomes:

$$\frac{\Delta H_{FP}}{L} \left(\frac{ft\, of\, fluid}{100\, ft\, of\, pipe} \right) = 1200\, f \frac{v^2 (ft/s)^2}{D(in) \times 2g (ft/s^2)} \qquad \text{[A-1]}$$

The friction factor is proportional to the Reynolds number which is defined as:

$$R_e = 7745.8 \frac{v(ft/s)\ D(in)}{v\ (cSt)} \qquad \text{[A-2}$$

The Reynolds number is proportional to the kinematic viscosity, the average velocity, and the pipe inside diameter. It is a non-dimensional number. The kinematic viscosity (v) is the ratio of the absolute viscosity (μ) to the fluid specific gravity (SG).

$$v(cSt) = \frac{\mu(cP)}{SG}$$

Flow regimes

Distinct flow regimes can be observed as the Reynolds number is varied from 0 to greater than 4000: Laminar flow, unstable flow, and turbulent flow.

Laminar flow - R_E < 2000

In the range of 0 to 2000, the flow is uniform and is said to be laminar. Laminar flow (also called streamline flow) occurs when a fluid flows in parallel layers, with no disruption between the layers. Looking at a longitudinal section of the pipe, the velocity of individual fluid particles is zero close to the wall and increases to a maximum value at the center of the pipe with every particle moving parallel to its neighbor. If we inject dye into the stream, we would notice that the dye particles maintain their cohesion for long distances from the injection point.

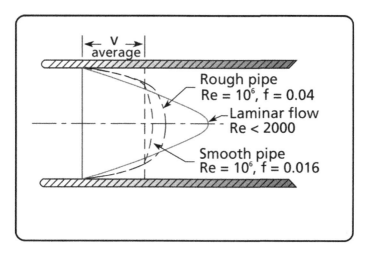

Figure A.1 Laminar and turbulent flow velocity profiles.

The friction loss is generated within the fluid itself. Figure A.1 shows that each fluid layer (in this case each fluid ring) of fluid is moving progressively faster as we get closer to the center. The difference in velocity between each fluid layer causes the friction loss.

The friction factor *f is* given by:

$$f = \frac{64}{R_e} \qquad \text{[A-3]}$$

For viscous fluids (i.e.: v ^{3} 50 SSU), the combination of velocity and viscosity usually produces a low Reynolds number and therefore laminar flow within a pipe. Pumping viscous fluids at a faster rate may cause the fluid to become turbulent, resulting in high friction losses. Equation [A-3] can be theoretically derived and is found in most fluid dynamic reference books. An interesting aspect of laminar flow is that pipe roughness is not a factor in determining friction loss.

Unstable flow - 2000 < R_E <4000

Unstable flow is characterized by its pulsing and unstable nature and possesses characteristics of both laminar and turbulent flow.

Turbulent flow - R_E > 4000

At Reynolds numbers larger than 4000, it is very difficult to predict the behavior of the fluid particles, as they are moving in many directions at once. If dye is injected into the stream, the dye particles are rapidly dispersed, demonstrating the complex nature of this type of flow. Reynolds, who originally did this experiment, used it to demonstrate the usefulness of a non-dimensional number (the Reynolds number) related to velocity and viscosity. Most industrial applications involve fluids in turbulent flow. The geometry of the wall (pipe roughness) becomes an important factor in predicting the friction loss.

Many empirical formulas for turbulent flow have been developed. Colebrook's equation is the one most widely accepted:

$$\frac{1}{\sqrt{f}} = -2\log_{10}\left(\frac{\varepsilon}{3.7D} + \frac{2.51}{R_e\sqrt{f}}\right) \qquad \text{[A-4]}$$

where ε is the average height of protuberances, called absolute roughness, of the pipe wall surface (for example, 0.00015 feet for smooth steel pipe). The term ε/D is called the pipe roughness parameter or the relative roughness. Since it is not possible to derive an explicit solution for f, L.F. Moody developed a graphical solution (see Figure A-4). The Moody diagram shows the linear relationship of the friction factor (f) with the Reynolds number (R_e) for the laminar flow regime. For Reynolds numbers in the medium range (4,000 to 1,000,000, turbulent flow), the friction factor is dependent on the Reynolds number and the pipe roughness parameter. This region is known as the transition zone. For high Reynolds numbers (1,000,000 and higher, i.e. fully turbulent flow), the friction factor is independent on the Reynolds number and is proportional only to the pipe roughness parameter. This region is the zone of complete turbulence.

Some typical values for the absolute roughness *e are provided below:*

Table A-1- Typical values for pipe wall roughness

PIPE MATERIAL	Absolute roughness e (feet)
Steel or wrought iron	0.00015
Asphalt-dipped cast iron	0.0004
Galvanized iron	0.0005

A numerical method, called the Newton-Raphson iteration technique, can also be used to solve the Colebrook equation. (Note: the Colebrook equation is valid only for Newtonian fluids.) Another equation for determining the friction factor (f) that was developed by Swamee and Jain, gives an explicit result for f and agrees with the Colebrook equation within 1%:

$$f = \frac{0.25}{\left(\log_{10} \left(\frac{\varepsilon}{3.7D} + \frac{5.74}{R_e^{0.9}} \right) \right)^2}$$

[A-5]

Determining Fittings Friction Head Losses for Newtonian Fluids using the K Method and 2K Method ($\Box H_{FF}$)
THE K METHOD
The fittings friction loss is given by:

$$\Delta H_{FF}(ft\ fluid) = K \frac{v^2 (ft/s)^2}{2g(ft/s^2)}$$

[A-6]

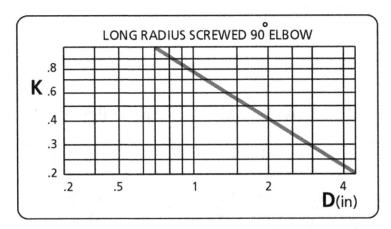

Figure A-2 Typical values for K with respect to fitting diameter.

The K factor for various piping fittings can he found in many publications, such as the "Pump Handbook" and the "Cameron Hydraulic Data Book". As an example, Figure A-2, depicts the relationship between the K factor of a 90° screwed elbow and the diameter (D). The type of fitting dictates the relationship between the friction loss and the pipe size. For example, the K value for a 2" 90° screwed elbow is 0.4.

Another good source for fitting K factors is the Crane Technical Data brochure.

STANDARD TEES

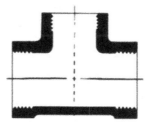

Flow thru run.......$K = 20\,f_T$
Flow thru branch....$K = 60\,f_T$

Figure A-3 Values for the K factor with respect to the friction parameter for a standard tee.

The Crane technical paper gives the K value for a fitting in terms of the term f_T as in this example for a standard tee (see Figure A-3).

$$K = 60 f_T \qquad \text{[A-7]}$$

As is the case for the data shown in Figure A-2, the friction loss for fittings is based on the assumption that the flow is highly turbulent; in fact that it is so turbulent that the Reynolds number is no longer a factor and pipe roughness is the main parameter affecting friction. This can be seen in the Moody diagram (see Figure A-4). There is a line in the diagram that locates the position where full turbulence starts.

The term f_T used by Crane is called the friction parameter and is the same parameter as that given by the Colebrook or the Swamee-Jain equation.

When the Reynolds number becomes large, the value of f_T (using the Swamee-Jain equation) becomes:

$$f = \frac{0.25}{\left(\log_{10}\left(\dfrac{\varepsilon}{3.7D} \right) \right)^2} \qquad \text{[A-8]}$$

Furthermore, the Crane data brochure assumes that the roughness of the material will correspond to new steel whose value is 0.00015 ft. Therefore, the previous equation for f_T becomes:

$$f = \frac{0.25}{\left(\log_{10}\left(\dfrac{0.00015 \times 12}{3.7D} \right) \right)^2} \qquad \text{[A-9]}$$

Therefore the value the K factor is easily calculated based on the diameter of the fitting, the friction parameter f_T and the multiplication factor for each type of fitting listed in the Crane data brochure.

THE TWO K METHOD

Tests conducted on various fittings have determined that the *K* value is not dependent on size, but on the Reynolds number. This approach takes into account the different nature of laminar and turbulent flow.

$$K = \frac{K_1}{\text{Re}} + K_\infty \qquad \text{[A-10]}$$

where K_1 and K_∞ are constantsare constants appropriate to the geometry of the fitting (see Table A-2). The examples in this book use the 2K method.

TABLE A-2 - *Typical values for K_1 and $K_¥$ for the 2K method*

FITTING TYPE	K_1	$K_¥$
90° ELBOWS		
Standard (R/D =1) screwed	800	0.4
Standard (R/D =1) flanged/welded	800	0.25
Long radius (R/D=1.5) all types	800	0.2
Mitered (R/D=1.5) 1 weld 90°	1000	1.15
Mitered (R/D=1.5) 2 weld 45°	800	0.35
Mitered (R/D=1.5) 3 weld 30°	800	0.3
Mitered (R/D=1.5) 4 weld 22 1/2°	800	0.27
Mitered (R/D=1.5) 5 weld 18°	800	0.25
45° ELBOWS		
Standard (R/D =1) all types	500	0.2
Long radius (R/D=1.5) all types	500	0.15
Mitered 1 weld 45°	500	0.25
Mitered 2 weld 22 1/2°	500	0.15
180° ELBOWS		
Standard (R/D =1) screwed	1000	0.6
Standard (R/D =1) flanged/welded	1000	0.35
Long radius (R/D=1.5) all types	1000	0.3
TEES, AS ELBOWS		
Standard screwed	500	0.7
Long radius screwed	800	0.4
Standard flange or welded	800	0.8
Stub-in branch type	1000	1
TEES, AS RUN-THROUGH		
Standard screwed	200	0.1

Long radius screwed	150	0.5
Stub-in branch type	100	0

VALVE: GATE, BALL OR PLUG		
Full line size b=1.0	300	0.1
Full line size b=0.9	500	0.15
Full line size b=0.8	1000	0.25
Globe standard	1500	4
Globe, angle or Y-type	1000	2
Diaphragm, dam type	1000	2
Butterfly	800	0.25

FITTING TYPE	K_1	$K_\yen$
CHECK VALVES		
Lift	2000	10
Swing	1500	1.5
Tilting Disk	1000	0.5

PIPE ENTRANCES AND EXITS		
Pipe entrance, normal	160	0.5
Pipe entrance, inward projecting	160	1
Pipe exit	0	1

Summation of Piping Losses

Recall that total system friction head, as defined here, is made up of the cumulative fluid energy losses due to the fluid movement and the friction loss due to the effect of pipe fittings (for example, 90° elbows, 45° bends, tees, etc.):

$$DH_F = DH_{FP} + DH_{FF}$$

The subscript FP used here refers to pipe friction loss and the subscript FF refers to fittings friction loss.

It is best that all the individual loss values be listed and labeled in a table so that they can be easily reviewed and summed. Remember that your total system friction loss corresponds to only one flow rate. If you wish to develop a system head curve, you will need to calculate total friction head losses for several flow cases.

The overall system curve will be composed of the static head requirements and the friction head losses at the flow rates of interest. Refer to the section "The Importance of System Head Curve" in Chapter 1 for more information on system head curves.

Figure A-4 The Moody diagram, friction factor vs. Reynolds number for laminar and turbulent flow at various pipe roughness-values.

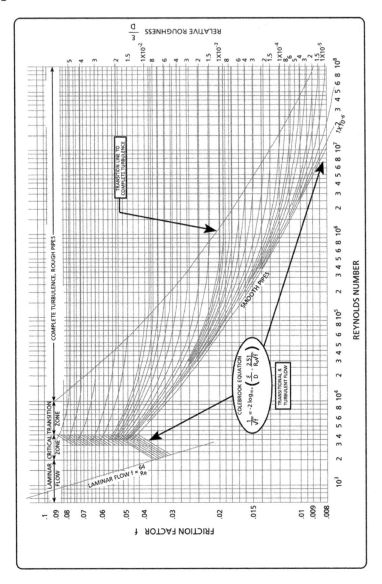

Some Illustrative Examples

Reynolds number example

Assume the following inputs:

D=4″

Velocity=12.77 ft/sec

Viscosity=1.0 cP

Solving for Re we get:

$$R_e = 7745.8\ \frac{v(ft/s)\ D(in)}{v\ (cSt)} = 7745.8 \times \frac{12.77 \times 4}{1.0} = 395,655$$

A Reynolds number of 395,655 means you are in the turbulent flow range.

THE K METHOD Example

Recall that the fittings friction loss is given by:

$$\Delta H_{FF}(ft\ fluid) = K\frac{v^{\,2}(ft/s)^{\,2}}{2g(ft/s^{\,2})} \qquad \text{[A-6]}$$

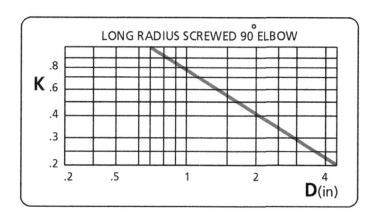

Figure A-2 Typical values for K with respect to fitting diameter.

Assume that we have a fluid velocity of 10 feet per second inside on a 4″ screwed elbow. By inspection of Figure A-2, we see that K equals about 0.22. Plugging in to equation A-6 we get:

$$\Delta H_{FF}(ft\ fluid)=K\frac{v^{2}(ft/s)^{2}}{2g(ft/s^{2})}=0.22\times\frac{10^{2}}{2\times32.2}=0.34\,ft$$

Moody Friction Factor Example

Recall that L.F. Moody developed a graphical solution for determining the pipe friction factor f as seen in Figure A-4.

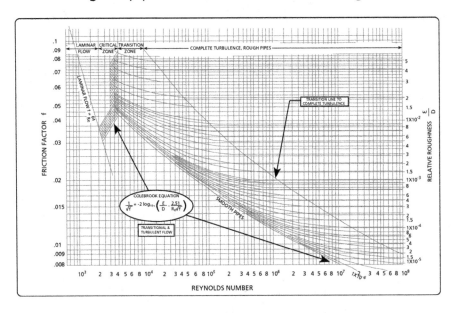

Figure A-4 The Moody diagram, friction factor vs. Reynolds number for laminar and turbulent flow at various pipe roughness-values.

Let's assume the results from the Reynolds number example above, where Re=395,655. If we assume we have a steel pipe, we see from Table A-1 that we have an Absolute roughness (ε (feet)) of 0.00015. This means that we have a relative roughness of 0.00015/4 = 3.75 x 10^{-5}.

Following the 3.75 x 10⁻⁵ line to the left until it passes the 395,655 value of Reynolds number on the Moody diagram we get a friction factor of about 0.016. Using the Darcy-Weisbach for resistance to movement of any fluid in a pipe, we get:

$$\frac{\Delta H_{FP}}{L} = f\frac{v^2}{D \times 2g} - 0.016 \times \frac{12.77^2}{4 \times 2 \times 32.2} = 0.01\,ft\,/\,ft$$

Or about 1 foot of loss per 100 feet of piping.

Appendix B

Mechanical Seal Selection Primer

A mechanical seal is an internal pump subassembly designed to prevent product leakage between a rotating shaft and the stationary parts housing. There are two basic categories of mechanical seals:

- Single mechanical seals (as shown in Figure B.1)
- Double mechanical seals (as shown in Figure B.2). Typically, double seals are used whenever the liquid being sealed poses significant safety and/or environmental risks.

Mechanical seals consist of two surfaces that slide against each other as the pump shaft rotates (see Figure B.1). The surfaces are held in place by either a spring or bellows arrangement and pressed together by the hydraulic pressure of the liquid. Between the two sealing surfaces, a fluid film is generated by the pumped product that prevents the mechanical seal face from touching the stationary ring. The absence of this fluid film will result in frictional heat and a rapid destruction of the mechanical seal. Some common conditions that can cause seals to fail prematurely are: a) running a pump dry, b) vaporization of the sealing liquid at the seal faces and c) excessive shaft vibration.

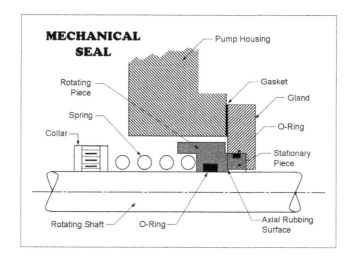

Figure B.1 – Typical Mechanical Seal

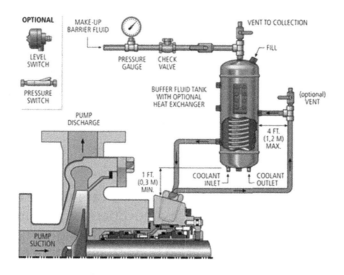

Figure B.2 – Centrifugal Pump with Double Seals

Mechanical seals can operate reliably for years as long as they've been properly selected for their application, installed properly, and operated with care. The person selecting the seal must clearly understand the anticipated operating conditions so that the right type of seal and materials of construction are chosen to achieve maximum seal performance. There are five key application factors you must understand in order to select the right mechanical seal for the application at hand:

Liquid Properties

Identifying the exact liquid being handled and its properties is the first step in the seal selection process. The seal material must be able to withstand the fluid being processed. In other words, all seal materials must be chemically compatible with the fluid at the actual pumping temperature. In addition, the liquid's vapor pressure must be known to ensure the fluid being sealed remains in a liquid state as it flows across the sealing faces.

Expected Seal Cavity Pressure

The pressure inside the seal chamber and the seal size determine the type of seal required and if it needs to be a balanced or unbalanced type seal. The pump manufacturer can estimate the expected pressure inside the seal cavity at normal pumping conditions. The seal operating pressure is a critical seal design parameter because it is used to determine if additional cooling is required to keep the liquid between the seal faces from flashing.

Sealing Temperature

Seal materials of construction must be carefully selected to handle the liquid at the anticipated operating temperature. It is vital to know a seal's maximum operating temperature because all seal materials have an upper temperature operating limit that should never be exceeded. Always provide the seal supplier with the full range of temperatures that the seal will experience. The design temperature range should even include a safety margin to cover process upset conditions.

Liquid Characteristics

It is vital to know the liquid's viscosity and solids content to ensure a reasonable seal life. Abrasive liquids can create accelerated wear that will ultimately shorten the seal's life. Double mechanical seals or an external seal flush plan can be used to cope with difficult fluids.

Reliability and Emission Concerns

Every plant has its own set of standards and operating procedures when it comes to mechanical seal reliability and emission requirements. The seal type and arrangement selected must meet or exceed the desired reliability and emission standards for the pump application.

After understanding the seal's expected operating conditions, we need to select the seal's overall materials of construction, including the face and component materials. When selecting the seal's material of construction, we need to consider the following characteristics of the material:

- Temperature constraints
- Chemical resistance properties
- Flexibility
- Wear resistance
- Thermal expansion and conductivity properties

When selecting the materials of the <u>seal faces and other internal components</u> materials for the mechanical seal, we need to consider these material characteristics:

- Wear resistance
- Low leakage & friction properties
- Good thermal properties
- Corrosion resistant

Single or Double Seal?

Single seals are the most common pump seal design employed due to their simplicity and their relatively low cost. However, regardless of their design and construction, a single-mechanical seal will eventually leak to the atmosphere, possibly leading to potential fire hazards, health hazards, or environmental concerns. Therefore, in hazardous or environmentally sensitive applications, it may be worthwhile to consider the use of a secondary or backup mechanical seal mounted within the pump's a seal chamber

Double seals can be designed in either tandem or double seal configurations. In a tandem seal, where two seals are oriented in the same direction, the process side seal leaks into a buffer fluid contained in the unpressurized cavity. A pumping ring or thermo-siphon pot is used to circulate buffer fluids in a circuit to promote cooling. (Figure B.7 (below) shows a typical API 52 flush piping plan.) If the buffer fluid cavity registers a dramatic increase in pressure or fluid level, the operator will know that the primary seal has failed. Seal leakage monitoring can be achieved by using either pressure or level switches or transmitters. If the buffer seal cavity liquid level is below normal or nonexistent, the secondary seal has failed. In both instances, maintenance will need to be performed on the seal.

Two advantages to a tandem design are that 1) there are two completely independent seals and 2) seals that are orientated in this configuration can withstand much higher pressures in the pump casing when compared to a double seal arrangement. Tandem seals are commonly used when the sealing liquid creates a hazard or changes state when contacting open air.

In a double seal, where two seals are oriented in opposite directions, there is a pressurized barrier liquid that is circulated in the cavity between the two seals. (Figure B.7 shows a typical API 53A flush piping plan.) The barrier liquid is circulated between the seals at a higher pressure than that of the process fluid so that if the primary seal fails, the barrier liquid will leak into the process instead of the hazardous pumped fluid escaping into the atmosphere. Double seal designs are usually used in unstable, highly toxic, abrasive, corrosive, and viscous fluids. One advantage of a pressurized double seal is that it allows the use of a less exotic inboard seal material so they are less expensive to purchase and maintain.

Seal Flush Plans

To obtain an acceptable seal lifetime, we must provide them with the best possible operating environment. For the more severe duty applications <u>seal flush plans</u> may be used to improve the operating environment of a seal. A seal flush plan is defined as any combination of accessories external to a seal with the purpose of improving seal life and/or reducing emissions. Seal flush plans are numerous and varied in design and purpose. First let's look at some of the common functions that seal flush plans can provide:

1. **Cooling**—All mechanical seals generate heat. To expel the generated heat, cooling must be provided either in the form of liquid circulation from the pump's discharge as provided by an API flush plan 11, shown in Figure B.3 with in a single seal arrangement.

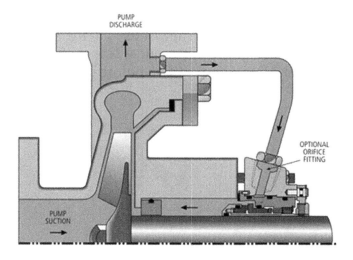

Figure B.3- API flush plan 11

Another means of providing cooling is the API 21 flush plan, shown in Figure B.4 in a single seal arrangement, which incorporates an inline water cooled heat exchanger.

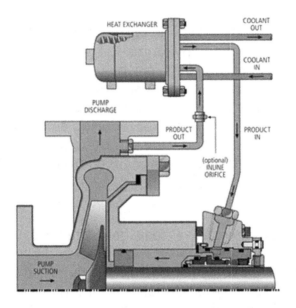

Figure B.4- API seal flush plan 21

2. **Filtering**—If a process stream has the potential of containing solids, a strainer, as used in the API flush plan 12, shown in Figure B.5 in a single seal configuration, can be employed to remove them before getting to the seal.

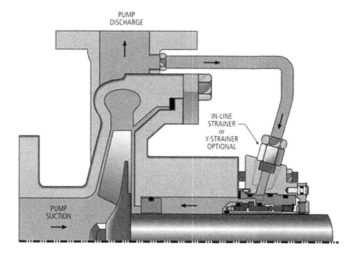

Figure B.5- API seal flush plan 12

3. **Isolation of the process side seal**—If the pumpage
   is judged to be undesirable for lubrication of the inner
   or process side seal due to its lubricity or cleanliness,
   an external flush can be used as shown in the API flush
   plan 32, shown in Figure B.6 in a single seal installation.
   Preferably a cool, clean liquid is used as a flush. Notice
   this seal flush design often incorporates a flow meter of
   some type as an aid to control and monitor flow.

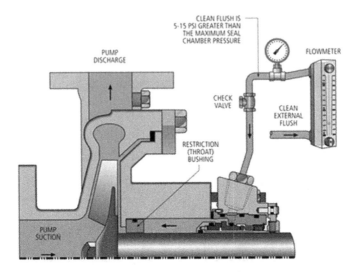

Figure B.6- API seal flush plan 32

4. **Isolation of the process from the atmosphere**—If external leakage of the pumpage is undesirable for environmental or safety reasons, a buffer fluid can be employed between the inner and our seal. This seal flush plan, called a plan 52, incorporates an unpressurized seal pot. Notice that Figure B.7 shows the API 52 flush plan being used with a combination inner and outer seal, also called a tandem seal. If the primary seal fails the outer seal will safely contain the pumpage until the pump can shut down and repaired. If installed, a level switch on the seal pot can alert the control room when the seal pot levels rises, due to an inner seal failure.

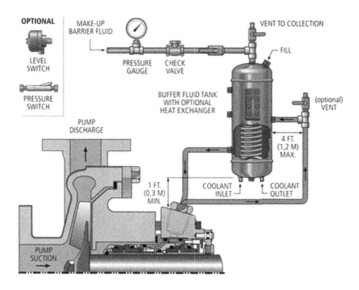

Figure B.7- API seal flush plan 52

5. **Isolation of the process side seal and the atmosphere**—In situations where you wish to avoid external leakage to the atmosphere as well as improve the inner seal's environment, an API seal flush plan 53A can be used. The key feature of this plan is the use of a pressurized gas blanket over the buffer fluid in the seal pot (see Figure B.8.) This plan requires a double seal be used as seen below.

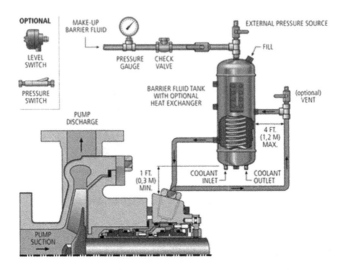

Figure B.8- API seal flush plan 53A

To ensure the proper operation of this seal plan, you must maintain a positive pressure differential between the seal pot pressure and the sealing chamber or stuffing box pressure. It is recommended that a differential pressure of at least 30 psi be maintained at all times to prevent flow reversal at the inner seal.

6. **Seal condition monitoring**—Seals with seal pots incorporating level glasses and/or pressure transmitters provide a means of tracking seal conditions (see API seal flush plan 52 above.) If the seal pot level drops, you probably have a leak in the outer seal and if seal pot levels rises, you probably have a leak in the inner seal. If you develop a massive leak, your pressure switch alarm will go off as you overfill the seal pot.

7. **Quenching**—Leakage that has a tendency of coking, setting-up, or crystallizing can be tamed with the use of an external quench flow, such as seen in the API 62 seal flush plan (see Figure B.8).

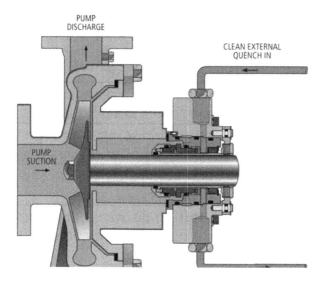

Figure B.9 API seal flush plan 62

You can find a listing of the most common seal flush plans in Table B.1. Work with the seal supplier to select the best seal flush design for you applications.

Table B.1-Common Mechanical Seal Flush Plans

API/ISO	ANSI	General Description
01	7301	Internal recirculation from pump discharge.
02	7302	Dead-ended, no circulation.
11	7311	By-pass from discharge to seal chamber.
12	7312	By-pass from discharge through strainer to seal chamber.
13	7313	Recirculation from seal chamber to pump suction.
14	7314	By-pass from discharge to seal and back to pump suction.
21	7321	By-pass from discharge through cooler to seal chamber.

22	7322	By-pass from discharge through strainer, orifice, cooler to seal chamber (not shown). Similar to Plan 21 with addition of a strainer.
23	7323	Recirculation from pumping ring through cooler to seal chamber.
31	7331	By-pass from discharge through cyclone separator to seal chamber.
32	7332	Injection from external source to seal chamber.
41	7341	By-pass from discharge through cyclone separator and cooler to seal chamber.
52	7352	Nonpressurized external reservoir with forced circulation.
53A	7353A	Pressurized external reservoir with forced circulation
53B	7353B	Pressurized external bladder type reservoir with forced circulation. Has been known as *Plan 53 Modified*.
53C	7353C	Pressurized external piston type reservoir with forced circulation.
54	7354	Circulation of clean fluid from external system.
61	7361	Tapped connections only. Usually used for Plan 62 to be installed at a later date.
62	7362	Quench fluid from external source.
65	N/A	Single seal leakage alarm for high leakage.
71	7371	Tapped connections only. Usually used for Plans 72, 75, 76 to be installed at a later date.
72	7372	External buffer gas purge for secondary containment seals.
74	7374	Pressurized external barrier gas for Dual Gas Seals.

| 75 | 7375 | Secondary containment seal drain for condensing leakage. |
| 76 | 7376 | Secondary containment seal drain for non-condensing leakage. |

Parting Advice:

As we have seen in this brief overview of mechanical seals, they are complex and varied in design and construction. Always keep in mind that mechanical seal design is a highly specialized endeavor. There are mechanical seal experts who have spent their whole lives studying seal design and learning the all the pertinent design codes. When in doubt, talk to the seal manufacturer before deciding on your final design. They have extensive experience in applying to mechanical seals to challenging sealing application across the industry.

Appendix C

Centrifugal Pump Selection Checklist

Date:
Pump Number:
Pump Name:
By:

The following checklist will assist you in gathering the required information for a centrifugal pump selection:

Preliminary Questions:

1. Do you have a finalized pump data sheet with liquid properties and process requirements?

    Yes _____ No _____

    If you answered "no" work with the process engineer to get a completed pump data sheet.

2. Have you conducted a thorough pumping system hydraulic analysis? A system head curve is preferred, but a single point is the minimum requirement for analysis.

    Yes _____ No _____

    If you answered "no", go back and analyze the pumping system in order to understand the pump hydraulic requirements.

3. Is there less than 2% entrained gas or vapor in the fluid to be pumped?

    Yes _____ No _____

4. Are solids in liquid small enough to pass through the pump impeller passages? As a rule of thumb the solid particles should be less than half of the smallest impeller opening.

Yes _____ No _____

5. Is the fluid viscosity less than 1000 cSt? If the viscosity is greater than 1000cSt, make sure to consider the economics related to the higher pumping energy cost.

Yes _____ No _____

If you answered "no" questions 3, 4, or 5, you should reconsider using a centrifugal pump.

Pump Hydraulics:

6. Calculate the Ns value using the information from the data sheet. Is the Ns value greater than 500 and less than 6000?

Yes _____ No _____

If the Ns is too low, consider increasing the number of stages or increasing the rotation speed. As a rule of thumb, you should consider 3600 rpm as the upper limit for most applications. Note: If you can't find a combination of impeller stages or speed that will increase the Ns value above 500, consider using a plunger, screw, or diaphragm pump.

7. Calculate the Nss value using data from the data sheet. Is the Nss value above 11,000?

Yes _____ No _____

If the Nss value is above 11,000, consider either increasing the NPSHa or using a vertical turbine pump. (Note: When using the Nss formula to calculate a preliminary Nss value, set NPSHr equal to NPSHa plus a healthy NPSH margin, i.e. NPSHr = NPSHa + NPSH margin.) A vertical turbine pump places the first stage impeller well below the suction flange to gain suction head. Another option is to decrease the rotational speed. Talk over the various options with the process engineer to see which one makes the most sense. If you discover there is no way to satisfy the Ns and Nss requirements, you may have to consider another type of pump.

Pump Selection:

8. Does your analysis indicate that your pumping requirements are well suited for a centrifugal pump selection?

   Yes _____ No _____

   If you answered yes you can proceed with requesting pump bids from qualified manufacturers. It is best to have at least 3 to 4 manufacturers bid if possible. The absolute minimum is two bidders to obtain competitive quotes. It is acceptable for manufacturers to offer more than one pump option. They will sometimes bid an alternative option to increase their chances of winning the bid.

9. Have you performed a detailed bid tabulation to enable you to objectively evaluate all the bids?

   Yes _____ No _____

   If you answered "no" you should take the time to perform a bid tabulation. It is best to list the top three selections in order of best to worst after you have completed the bid analysis. If for some reason your first choice drops out or is disqualified, you will have your other options identified.

10. Have you sat down with the process engineer, maintenance engineer, the purchasing agent, and any others with vested interests, to review the bid tabulation?

    Yes _____ No _____

    If you answered "no" you should set up a meeting to get everyone's input and get a consensus on the final pump selection.

Appendix D

Useful Conversions

When analyzing and evaluating pump information, it is critical that you carefully document your measurements units to ensure dimensional consistency. For example, if the manufacturer recommends a maximum bearing temperature of 175 °F, you must ensure that you either record the reading in degrees Fahrenheit or you must convert your Celsius reading into degrees Fahrenheit for a direct comparison. Similarly, if you are measuring pressure in psig units in the field, but the manufacture has provided pressure in kilopascals, you must be able to convert psi units into kilopascals units in order to properly compare the manufacturer's performance information with your data.

Always ask the following questions to ensure you are using the proper measurements units in your evaluation:

- What are the units of the pump data that I am gathering in the field?
- What are the corresponding measurement units provided by the pump manufacturer?
- Do I need to make a unit conversion to compare my readings with the manufacturer's performance data? If you do have to convert the readings to the using a conversion factor, make sure you are consistent throughout your analysis.

If you are not sure about your measurement units, ask someone who can help.

Here are some conversion formulas and tables you might find useful when dealing with machinery:

Rotational Speed Conversions

Revolutions per minute = Revolutions per second x 60

Example #1: 60 revolutions per second x 60 = 3600 revolutions per minute

Rotations per second = Rotations per minute / 60

Example #2: 1800 revolutions per minute / 60 = 30 revolutions per second

Speed

Inches per second = Millimeters per second / 25.4
Millimeter per second = Inches per second x 25.4
Inches per second = Feet per second / 12
Feet per second = Inches per second x 12
Inches per second = Meters per second x 39.37
Meters per second = Inches per second /39.37

Distance and Length Conversions

Metric Conversion Factors	
1 centimeter =	10 millimeters
1 decimeter =	10 centimeters
1 meter =	10 decimeters
1 Angstrom =	10^{-10} meters

English & Metric Conversion Factors	
1 inch =	25.4 millimeter
1 inch=	25,400 microns
1 feet =	12 inches
1 yard =	3 foot
1 mil =	0.001 inch

Common Volume & Capacity Conversion Factors

1 cubic yard (cu yd.) = 27 cubic feet
1 teaspoon = 1/3 tablespoon
1 tablespoon = 1/2 fluid ounce = 3 teaspoons
1 U.S. fluid ounce (fl oz.) = 1/128 U.S. gallon = 1/16 U.S. pint

1 cup = ¼ quart = ½ pint = 8 fluid ounces
1 pint (pt.) = 1/8 gallon = 1/2 quart = 16 fluid ounces
1 quart (qt) = 1/4 gallon = 32 fluid ounces
1 U.S. gallon (gal) = 231 cubic inches
1 milliliter (ml) = 1/1000 liter = 1 cubic centimeter
1 centiliter (cl) = 1/100 liter = 10 milliliters
1 deciliter (dl) = 1/10 liter
1 liter = 1 cubic decimeter
1 cubic inch = 16.4 cubic centimeters
1 cubic foot = 0.0283 cubic meter
1 cubic yard = 0.765 cubic meter
1 fluid ounce = 29.6 milliliters
1 U.S. pint = 0.473 liter
1 U.S. quart = 0.946 liter
1 U.S. gallon = 0.84 imperial gallon =3.8 liters
1 dry pint = 0.55 liters
1 dry quart = 1.1 liters
1 cubic centimeter = 0.06 cubic inch
1 milliliter = 0.034 fluid ounce
1 liter = 1.06 U.S. quarts = 0.9 dry quart

Temperature Conversions

Fahrenheit to Centigrade temperature conversion: °F = (1.8 x °C) +32

Centigrade to Fahrenheit temperature conversion: °C = (°F-32) x 0.555

> Example #1: 212 °F = (212 -32) x 0.555 = 100 °C
> Example #2: 130 °C = (1.8 x 130) + 32 = 266 °F

Vibration Conversions

Inches per second = millimeters per second / 25.4
Inches per second = microns per second / 25400
Millimeter per second = inches per second x 25.4
Microns per second = inches per second x 25.400
Table C.1- Common English to Metric Conversions

Common English to Metric Conversions

Type of unit	Multiply This known value		by this conversion factor	to get this desired value
length	inches	X	25.4	millimeters
length	feet	X	0.3	meters
area	square inches	X	6.5	square centimeters
area	square feet	X	0.09	square meters
volume	cubic inches	X	16	cubic centimeters
volume	cubic feet	X	0.03	cubic meters
volume	quart	X	0.9	liter
volume	gal	X	0.004	cubic meters
weight	ounce	X	28.3	gram
weight	pound	X	0.45	kilogram
power	horsepower	X	0.75	kilowatt
speed	feet per second	X	0.304	meters per second
force	ounce force	X	0.278	newtons
force	pound force	X	4.448	newtons
work/torque	foot pounds	X	1.355	newton meters
work/torque	inch pounds	X	0.112	newton meters
frequency	cycles per second	X	1	hertz
pressure	pounds per square inch	X	6.897	kilopascals
pressure	pounds per square inch	X	0.0069	megapascals
pressure	pounds per square inch	X	0.07	kilogram per square centimeter
density	Pounds per cubic foot	X	16.02	kilograms per cubic meter
density	Pounds per cubic inch	X	27.68	grams per cubic meter

density	Pounds per cubic inch	X	27680	kilograms per cubic meter
heat	British thermal units	X	1055	joules

Common conversion examples:

Example #1: You have recorded a pressure of 1000 psi. What is the pressure in megapascals?

Solution: 1000 psi x 0.0069 = 6.9 megapascals

Example #2: You have an electric motor rated at 100 Hp. What is the kw rating of the motor?

Solution: 100 hp x .75 = 75 kilowatts

References:

1. Karrassik, et. al, "Pump Handbook, "McGraw-Hill, Third Edition, 2001
2. Perez, R. X., "Operators Guide to Centrifugal Pumps," Xlibris, 2008, pp15-16
3. Yedidiah, S., Centrifugal Pump User's Guidebook, 1996.
4. Lobanoff, V.S. and Ross, R.R., Centrifugal Pumps: Design and Application, second edition, 1992.
5. Hallam, J. L., "Centrifugal Pumps: Which Suction Specific Speeds are Acceptable?", Hydrocarbon Processing, April 1982.
6. Perez, R. X. and Stark, W. B., "Financial Calculations for Pumps," Pumps and Systems Magazine, May 2007
7. Chaurette, Jacques, "Pump System Analysis and Sizing," 5th Edition February 2003. Printed in Canada, Published by Fluide Design Inc,, 5764 Monkland avenue, Suite 311, Montreal, Quebec.

Index

O

Open (atmospheric) pit   35

P

PIP guideline RECP001   83
Pipe wall roughness   93
Piping system   14-17, 21-2, 30, 36, 89
Pressure head   5-6
Pressurized system   19, 30
Pressurized tank   33
Process engineer   45-6, 88
Project engineer   45-6
Pump and system interaction   22
Pump curve   7, 17, 21-2, 37, 46, 48, 76-7
Pump design analysis   57, 60-1
Pump Handbook   19
Pump impeller   2, 24-5, 37, 43, 48, 119
Pump selection philosophies   46
Pump selection process   45, 47, 63, 85
Pump selection rules of thumb   65
Pump specification sheet   86
Pump supplier   45-6
Pumping hydrocarbons   38
Pumping system   15, 20, 38-9

Q

Quenching   114

R

Reliability and emission concerns   107

Reynolds number   90, 92-3, 101-3
Rotating equipment engineer   45-7
Rotational speed conversions   122
Rule of thumb   19, 25, 51, 69, 119

S

Seal cavity pressure   106
Seal condition monitoring   114
Seal flush plans   109, 115
Sealing temperature   106
Shaft   9, 25, 65, 104
Single mechanical seal   104
Single or Double Seal?   107
Solids   9, 24-5, 43, 87, 106, 110, 119
Specific gravity   3-5, 18, 33-5, 49, 55, 57, 60-1, 79-80, 90
Specific speed   xii, 22, 39, 41-3, 46, 48, 50-1, 54-5, 57, 59, 67-72, 75-6
Speed conversions   122
Steam turbine   13
Strainers   24
Suction energy   54-5, 58, 61-2, 72
Suction lift   30-1
Suction specific speed   xii, 39, 46, 48, 51, 54-5, 57, 67, 71-2, 75-6
Swamee-Jain equation   95
System head   15-18, 20-2, 99, 118
System head curve   15-17, 21, 99, 118

Printed in the United States
By Bookmasters